Praise for *The Optimist's Telescope*

"How might we mitigate losses caused by shortsightedness? Bina Venkataraman, a former climate adviser to the Obama administration, brings a storyteller's eye to this question. . . . She is also deeply informed. . . . Heed Venkataraman's impassioned call for making a commitment to future change."
—*The New York Times Book Review*

"Bina Venkataraman zeroes in on the heart of [the] problem. . . . With vivid anecdotes, [she] musters all the optimism she can manage to suggest how we might be able to overcome something that feels like plain old human nature."
—NPR, Best Books of 2019

"Venkataraman vividly depicts what happens when we don't plan ahead and what we can do about it, on our own and together." —*The Washington Post*

"In pacy prose that's easy to digest but offers concrete examples of change . . . this book offers hope that we can take back some control of our own destinies and aim for a better future." —*Financial Times*

"Most of us only see the future after it becomes the past. *The Optimist's Telescope* is here to change that. It's a rare read that's as fascinating as it is important. In it, Bina Venkataraman brings together powerful narrative, cutting-edge behavioral science, and the rich experience of a high-impact career."
—Adam Grant, author of *Originals*

"The unknown can always be scary. But in this wise, eye-opening, and hopeful book, Bina Venkataraman shows us the ways we can think more clearly and strategically about the future—in our communities, our families, and in our own lives."
—Arianna Huffington

"Bina Venkataraman illuminates how we can make better decisions for ourselves, our communities, and Earth itself. She introduces us to an array of colorful and unexpected characters, from ancient philosophers to tech entrepreneurs, all while distilling the science of foresight into practical advice we can all use. A timely and valuable book." —Anne-Marie Slaughter, author of *Unfinished Business*

"*The Optimist's Telescope* is a noble and important book. Through stories of people who have made a difference and an acute awareness of how things can be made better, Bina Venkataraman shows how we can effect change and make the world a better place. She is the good parent this planet so desperately needs."
—Errol Morris, Oscar-winning filmmaker and author of *Believing Is Seeing*

"[Venkataraman] explores all sorts of ways that businesses, governments, and communities have learned to be better planners-ahead. . . . She also writes beautifully about how we can all 'be better ancestors.'" —*The Grist*

"A timely reminder that time is not on our side without long-term thinking."
—*Kirkus Reviews*

"A thought-provoking and eminently readable debut . . . Venkataraman's thoughtful and clear-eyed assessment of how to teach oneself to make more carefully considered decisions should prove a valuable tool for anyone wishing to think less in the short term and more toward the future." —*Publishers Weekly*

"An intriguing look at strategies for the long term with citations from business executives, sociologists, and philosophers; highly recommended."
—*Library Journal*

"Chinese peasants once got paid by the piece to find dinosaur bones; soon they took to smashing the bones into tiny pieces to boost their income. This book is a sobering compendium of the many ways in which ill-conceived short-term incentives undermine valuable long-term goals. The stakes go way beyond busted fossils: think rising sea waters, treatment-resistant pathogens, collapsing infrastructure, disappearing topsoil . . . A grim list and grounds for pessimism unless, that is, it gets read in the context of this book, for Bina Venkataraman has assembled a remarkable repair kit, full of tested tools for harmonizing the lure of present reward with the foresight we need if we are to build a durable future." —Lewis Hyde

"*The Optimist's Telescope* will help you think about the biggest decisions you face in your life and that humanity faces in this historical moment. Everyone should read this book." —Gary Knell, chairman of National Geographic Partners

THE OPTIMIST'S TELESCOPE

Thinking Ahead in a Reckless Age

Bina Venkataraman

RIVERHEAD BOOKS.NEW YORK

RIVERHEAD BOOKS
An imprint of Penguin Random House LLC
penguinrandomhouse.com

A version of the story about the dilruba handed down from the author's great-
grandfather (chapter 9) appeared in *New America Weekly* and *Pacific Standard*.

Haiku of Issa's ["In this world /
We walk on the roof of hell / Gazing at flowers"] from:
The Essential Haiku: Versions of Basho, Buson & Issa, edited and with
an introduction by Robert Hass. Introduction and selection copyright
© 1994 by Robert Hass. Reprinted by permission of HarperCollins Publishers.
Robert Hass, *The Essential Haiku: Versions of Basho, Buson and Issa* (Ecco, 2013;
Bloodaxe Books, 2013). Reproduced with permission of
Bloodaxe Books. www.bloodaxebooks.com.

The Library of Congress has catalogued the Riverhead
hardcover edition as follows:

Names: Venkataraman, Bina, author.
Title: The optimist's telescope : thinking ahead in a reckless age /
Bina Venkataraman.
Description: New York : Riverhead Books, 2019. |
Includes bibliographical references and index.
Identifiers: LCCN 2018046076 (print) | LCCN 2018046851 (ebook) |
ISBN 9780735219496 (ebook) | ISBN 9780735219472 (hardcover)
Subjects: LCSH: Decision making—Social aspects. |
Risk—Sociological aspects. | Forecasting.
Classification: LCC BF448 (ebook) | LCC BF448 .V46 2019 (print) |
DDC 153.8/3—dc23
LC record available at https://lccn.loc.gov/2018046076

First Riverhead hardcover edition: August 2019
First Riverhead trade paperback edition: August 2020
Riverhead trade paperback ISBN: 9780735219489

Printed in the United States of America
1 3 5 7 9 10 8 6 4 2

BOOK DESIGN BY AMANDA DEWEY

For my parents,
who crossed oceans for the sake of the future

CONTENTS

Not everything that is faced can be changed, but nothing can be changed until it is faced.

—JAMES BALDWIN

That's a problem for future Homer. Man, I don't envy that guy.

—HOMER SIMPSON

PREFACE

One summer morning several years ago, I went for a hike in the Hudson Valley, an idyllic landscape of wooded hills and wide pastures along the Hudson River just north of New York City. A trail took me through a meadow where I spotted a red-tailed hawk soaring overhead, and past a waterfall whose banks teemed with ferns. The next day, I returned to Washington, D.C., where I was living, and noticed a rash on the back of my leg. It was itchy and crimson, and a bump at the center looked like a spider bite.

I snapped a photo with my phone and made a mental note to get the rash checked by a doctor. At the time, I was working sixteen-hour days and rushing from deadline to deadline on little sleep: As silly as it might sound, the thought of taking an hour to go to a clinic seemed out of the question. Weeks went on, and the rash faded from my skin and my memory.

Eight months later, my knee swelled to the size of a grapefruit, and I discovered I had a bad case of Lyme disease that would take months to treat. I spent a brutal winter in Boston on crutches. I injected myself with a daily IV drip of medication until I finally got better. The tick that had bitten me that day in the woods had not left a telltale bull's-eye—but I actually knew that this was common. Years before, as a science journalist, I had written articles about the spread of Lyme disease. I knew that the Hudson Valley counties had the highest infection rates in the country.

Why had I chosen to ignore the warning signs? As I recovered, the question haunted me. The minor inconvenience of seeing a doctor right after my hiking trip—to treat the infection before it spread—could have spared me months of pain, and what turned out to be permanent knee damage. Or I could have just taken the precaution of wearing bug repellent and protective clothing.

As much as I knew about the disease, however, I just couldn't see myself ever getting it. I had gone on hundreds of hikes and had never had a tick bite, let alone any kind of infection. Outdoors, in my teens and twenties, I had been something of a daredevil—jumping off seaside cliffs, climbing trees, scaling steep rock faces. I suffered from delusions of invincibility.

I know I'm not alone in having made a mistake like this. Nor am I alone in feeling the subsequent regret. Anyone who has ever tried to warn a friend against dating someone bound to break her heart—or told a teenage driver not to speed—knows the pattern. It's a curious glitch in human behavior: Smart people make reckless decisions, despite clear warnings.

This conundrum is something I have faced not just in my personal life but also in my work. In one way or the other, for the past fifteen years, I've tried to warn people about threats that lie ahead. When I was a journalist, I wrote articles that alerted the public to the coming dangers of pandemics and wildfires, of what would happen to farms when reservoirs dried up. I teach college students at MIT how to do what I do—share scientific information with wider audiences for the sake of helping them make better decisions about the future.

The stakes got especially high when I went to work in the White House in 2013, where my job was to convince mayors, business leaders, and homeowners that they should get ready for a future of rising seas and deadly heat waves, of persistent droughts and biblical floods.

As I did this work, I often found myself failing to persuade people to take action. I got stuck in the same place over and over, as if I were running into a traffic jam no matter the route I took: Like me, most people didn't think much about distant consequences when making decisions. Families bought houses because the mortgage payments were low today, not weighing the downstream cost of flood damage. Food company executives boosted immediate profits for their shareholders rather than investing to protect farms from later droughts. Politicians, of course, spent far more energy plotting reelection that year than protecting their city or state for the next ten years.

I couldn't really blame any of them, because their decisions made perfect sense to them at the time. And I knew I wasn't any better at heeding warning signs myself.

For decades, I've watched communities learn about perils the hard way—by living through them. Only when severe droughts struck California and Cape Town, South Africa, and floods submerged Houston and Mumbai did residents and leaders in those places take the costs of such foreseeable disasters seriously. It was not just a matter of heeding warnings but of people, businesses, and communities being able to deeply consider the future consequences of the decisions they made every day.

We live in an era when, more than ever, we need to make smart choices for the future—for ourselves and for future generations. We are living far longer than our grandparents, our life spans waxing beyond our imaginations and retirement plans. We have tools to edit the traits of human embryos and to build intelligent machines—technologies that will redefine what it is to be human for coming generations. And we're shaping the planet's weather patterns for the next century in ways that may well destroy crops, drown coastal cities, and displace millions of people. To stave off deadly pandemics or stop the worst of climate change, we have to place a lot more value on what happens in our own future and that of people distant from us in time.

I call ours a reckless age—but not because we are somehow worse or weaker than our predecessors. Rather, the need has never been greater for a civilization to think ahead, because the stakes have never been higher. A world population of seven billion that has the power to put robots on Mars and invent new species can shape the future of humanity on an unprecedented scale, and with longer-lasting impact. At the same time, we have unparalleled knowledge—the ability to detect the warning signs of disaster more clearly than our forebears and to see the legacy of our choices—whether it's the half-life of radioactive waste or how pollution today kills off coral reefs tomorrow. The people who died in ancient Pompeii, by contrast, were not much wiser to danger than the dinosaurs that were wiped out by the Chicxulub meteorite. When you don't see catastrophe coming, you can't be called reckless. That's just misfortune.

In our lives, each of us has witnessed people making shortsighted decisions. A voter stays home on Election Day because his laundry seems more urgent than standing in line at the polls, then regrets not casting a ballot. A doctor prescribes painkillers to make a patient feel better right away, which soon leads to an opioid addiction. A married woman indulges in an affair that for decades afterward she wishes never happened. An executive slashes a research budget for new products and later finds the company failing. A family builds a dream house on a barrier beach, and in a few years, the home has been washed away.

What's more troubling are the reckless decisions we make in society. Early-warning signs of calamity—take the COVID-19 pandemic of 2020 or the 2007 subprime mortgage crisis in the United States—go neglected until it's too late. People in the richest countries save less for the future than those in poorer generations before them. Soaring stock market returns placate investors, masking trends likely to stall the global economy in the future. A person who tweets to forge foreign policy doctrine can be elected president of the United States, in part because what should be ephemeral insults command news cycles and our attention.

It is tempting to shrug this off as our doomed fate—to buy into the cynical view that such myopia is hopelessly ingrained in human nature, in our economy, and in society. But that view gravely underestimates our potential and conveniently excuses neglect of our obligations. It rests on the false assumption that it is impossible to think ahead. We know, however, that sometimes people, businesses, and communities *do* avert crisis and act for the sake of the future; they have throughout history, and they do even today. Civilizations have built pyramids and grand cathedrals, stopped ozone destruction and prevented nuclear Armageddon. Societies have educated generations of the poor, eradicated polio, and sent people to walk on the moon. What has made their foresight possible where others have failed?

My curiosity about that question led me to this book. For seven years and counting, I have been investigating what allows wisdom to prevail over recklessness; what role our biological programming, our environment, and our culture play; and what changes are possible in our communities, businesses, and society. I've taken my inquiry to dive bars, to city council meetings, to old-growth forests, to family reunions, and on foreign delegations around the world. I've visited Kansas farms, Wall Street firms, virtual reality labs in Silicon Valley, fishing villages in Mexico, and the nuclear fallout zone in Fukushima, Japan.

I've met people who have faced dilemmas similar to my own: A doctor trying to stop the rise of deadly superbugs. An investor proving his future prospects amid current losses. A community leader fighting a reckless real estate development. A police official warning of a looming terrorist attack. A farmer trying to prevent the next Dust Bowl. Each of them, like me, has tried to help others heed warnings before it's too late. Most want to do something bigger than themselves, to make the future better for other people. Their failures and triumphs hold lessons for the rest of us.

Along the way, I have also mined research findings from a wide range of disciplines and tapped experts in fields as diverse as archaeology, land-use law, engineering, economics, and evolutionary biology—from a social movements guru to an artificial intelligence mastermind, from a clock-maker to the U.S. secretary of defense. I've sought to understand what today's leading thinkers believe and what science and history tell us about how to better weigh future consequences.

What I have discovered is that the way I—and most people—have thought about our recklessness is wrong. What we believe is inevitable in human nature and society is rather a choice we face. As with the prisoners in Plato's cave allegory, whose chains prevented them from perceiving the real source of the shadows they saw, our perch has constrained our view of the possible. It's now time to find a path out of the cave.

I wrote this book to impart what I have learned about the untapped power we have, individually and collectively, to avoid the reckless decisions that put our future at stake.

A NOTE ON SOURCES

No single genre of evidence holds a monopoly on truth. In this book, I draw on many styles of source and many breeds of authority: research findings from randomized controlled trials and natural experiments, the practices of people and communities, and the wisdom of poets and philosophers. I triangulate truth with firsthand reportage, peer-reviewed science, expert opinion, historical precedent, and ancient intuitions.

Within the category of behavioral science research, I describe specific studies only when I have been persuaded by a larger body of research that backs up the use of the particular—or I offer caveats within the text or endnotes, inviting the reader's scrutiny.

I draw on the time-honored practice of telling stories. Although no single story can serve as ultimate or universal truth, I believe the narratives here reveal important aspects of who we are and what we face.

THE TROUBLE
WITH THE FUTURE

Let's say today you are going to the supermarket. When you arrive, you head to the produce aisle—the land of good intentions. Maybe you pick up some bananas, oranges, and leafy greens, the kind of food that makes you feel like you just took a shower or donated to a homeless shelter. You feel virtuous, like an upright citizen.

As you go deeper into the grocery store, however, you pass a devilishly attractive bag of potato chips. Your doctor has told you to avoid this stuff, but you're craving it. Looking at the bag, you can almost taste the salt and feel the crunch between your teeth. You start to salivate in aisle 7. This might not be the best choice for the future you but right now you want it. You toss the bag in your cart.

After picking up a few more items from your list, you make it to the checkout line. Behind the cashier, you spot some lottery tickets. You had planned to hoard your change this year to build up your savings, but playing the Diamond Dazzler costs only a couple of dollars. Who knows, you might even win. You pick up a few scratch tickets. In the future you might regret this, but for now the little indulgence feels good. You pay the cashier and make your way home.

Every day, each of us faces decisions where we must choose between what we want right away and what's best for us—and others—in the long run.

We decide whether to splurge on dining out or save for a rainy day, whether to bike to work or drive on a cold morning, whether to clean out a plastic container to recycle it or toss it in the trash. What we choose can vary depending on how desperate or sated we feel, how rushed or calm, how tired or energized, and even how optimistic. Maybe on some trips to the supermarket you resist your cravings, and on some you indulge. Once in a while, I find it's important to eat chocolate like there's no tomorrow.

Some choices between the present and the future, like my chocolate fix, are trivial; the decision likely won't determine whether I live or die or even my ultimate happiness. Such decisions can hardly be called reckless, even if we regret them—they are simply human.

But other choices matter. Maybe you want to take a trip next year or be more ready for emergencies, but you can't stop spending your savings on impulse purchases. Or perhaps you wish you could study harder now to learn a new language or get a degree that would open up doors later. In trying to avoid immediate pain or inconvenience, we often sacrifice our own future aspirations. Even choices that are inconsequential in isolation, like neglecting to exercise one day, can add up to eventual devastation, like a heart attack. A trifling post on social media, made in a rash moment, might do permanent damage to a career or reputation.

The decisions we make today can define our experience of the future. This is true for individuals, and also for businesses, communities, and societies. The power of present choices to shape what's to come reverberates in the stories of people I have encountered all over the world and in the annals of history: A poker player who took calculated risks to earn hundreds of thousands of dollars but whose father squandered the family savings at racetracks. Texas fishermen whose decisions led to the near extinction of the Gulf of Mexico's red snapper population, and then brought it back. Security officials who failed to protect the athletes at the

Munich Olympics from a terrorist attack, and thousand-year-old stone markers that saved Japanese villages from modern destruction. A Mississippi man who clung to life on a rooftop because he chose not to evacuate before Hurricane Katrina, and an Oregon teacher who trained herself not to reflexively punish students with dark skin. The philosopher who reassured the residents of ancient Pompeii that their city was safe, and the team that defused the Cuban Missile Crisis and prevented nuclear war.

Decisions made by societies carry the longest legacies: The ancient choice to grow annual crops in the Middle East had a hand in the American Dust Bowl and set the stage for the current loss of fertile lands around the world. Building the U.S. interstate highway system defined how people traveled and commuted for generations, and free high school education baked in the country's economic growth in the twentieth century.

This book is about the decisions we make as individuals and collectively that have great consequences for our lives and the lives of others, the decisions we may come to regret or celebrate. It is especially about decisions that are reckless—when we ignore clear signs of opportunity or danger in the future. Through close investigation of such decisions across many contexts, I have discovered the untapped power we have to make wiser choices.

Decisions involve both information and judgment—whether we make them alone or in groups. The judgment to make smart choices about the future is what I call foresight. To exercise foresight is different from seeing the future like the mythical prophet Cassandra, said to have predicted the fall of Troy. Volumes of research projects and books have been devoted to mimicking her clairvoyance—or at least to more accurately predicting the future. But too few have helped us develop the judgment we need to make decisions about the future. To exercise foresight is to weigh what we know—and also what we don't know—about what lies ahead, making the best call not just for the present but for the sake of our future selves. It's the difference between knowing it will rain at tomorrow's soccer game and actually bringing an umbrella.

I argue in this book that many decisions are made in the presence of information about future consequences but in the absence of good judgment. We try too hard to know the exact future and do too little to be ready for its many possibilities. The result is an epidemic of recklessness, a colossal failure to plan ahead. To correct course, we need to hone our foresight.

Many people today want to act on behalf of the future more than we actually do. We long to think past the fleeting instant of a text message, and for our lives to have meaning as a stitch in the long, intricate fabric of time. We aspire to do right by future generations and to be viewed by them with admiration, or at least without disgust. We suspect that if we could learn to think ahead, we might have more money, live healthier, and better protect our families from danger. Businesses could earn more profits, communities could thrive, and civilizations could avoid foreseeable catastrophes. We might even better steward forests, rivers, and oceans for posterity.

Yet people today are struggling to weigh future consequences, whether in our daily lives or in humanity's highest endeavors. It is hard to sacrifice for a delayed reward and easy to indulge now, even if it means courting later disaster. The more distant the consequences of our decisions, the more difficult it becomes to exercise wisdom about them.

Acting for the sake of the future is easiest when we do it for ourselves and it does not require much immediate sacrifice. Brushing your teeth twice a day is a small price to pay now to prevent a root canal or dentures. Writing a will might take a few hours of your time once every few years, but concern for your family overshadows the inconvenience. The more time and money you have, the easier it is to plan ahead, for instance, by buying health insurance or helping your kids with homework. The more control and certainty you have that your choices will make a difference, moreover, the more likely you are to act for the future.

When we're looking forward to something—a picnic with friends, a vacation, a wedding day—most people find it easier to imagine the future. We want to see ourselves in that moment, and so we let our minds wander there when we have the time. But when we dread something— doing our taxes, getting older, the rising seas, or a coming refugee crisis—most of us don't want to inhabit the future. Even if we fixate on it, we often find it anxiety-inducing or even paralyzing, because we wish it were not coming at all.

When it comes to making sacrifices now to get ready for eventual earthquakes in a community, to invest in future inventions for an industry, or to prevent overharvesting fish as a society, choosing the future is especially difficult. To act on behalf of our future selves can be hard enough; to act on behalf of future neighbors, communities, countries, or the planet can seem impossible, even if we aspire to that ideal. By contrast, it is far easier to respond to an immediate threat. This helps explain why, for instance, countries of the world failed to prevent the Ebola epidemic of 2014 that eventually killed more than ten thousand people—when it would have cost millions of dollars less to invest in vaccine research and medical facilities than it did to react to the deadly outbreak after it emerged.

But why, exactly, is it so hard to act for the sake of the future, even when we hope to make it better?

For one, we can't smell, touch, or hear the future. The future is an idea we have to conjure in our minds, not something that we perceive with our senses. What we want today, by contrast, we can often feel in our guts as a craving.

Temptations that we take in with our senses burn inside us with a kind of emotional heat, Walter Mischel, the late renowned Stanford psychology professor, argued. Smelling the fresh doughnut in the bakery case or seeing the colorful scratch ticket at the gas station register creates a hot feeling that makes it hard to weigh future consequences—even if in cooler moments, away from such temptations, we would choose to forgo a quick fix for the sake of the future.

What lies in the future—our health in middle or old age or eventual financial stability, a community that someday has cleaner water or safer streets—is murky, mutable, uncertain. It holds none of the surefire satisfaction of the plate of french fries on the diner counter. We rarely know for sure that giving up something today will yield what we want tomorrow.

It does not help that our future selves are strangers to us. Most people don't know what they'll want for dinner next Tuesday, so how could they know exactly what they will want in the next decade? Future generations are even more alien. We experience societal change today at a dizzying pace, with technological progress that quickly renders the future unrecognizable to the past. In the 1960s, the futurist Alvin Toffler had the prescience to identify this trend and its destructive effects on people's foresight, calling it "future shock." The malaise has only become more acute in the twenty-first century as we undergo more frequent upheavals in how we communicate, travel, and work. To make decisions for a world decades in the future while living in today's world can feel abstract and even fruitless.

For aeons, thinkers have contemplated why we sabotage our future selves when we make decisions—even when we have knowledge about the likely consequences. Aristotle wrote that *akrasia*, a weakness of human will, prevents us from having lives of meaning. But he also thought it ridiculous to strive for perfection in resisting indulgence, which at times brings us sustenance and pleasure. Better to strike a balance, he said, avoiding rash decisions through practices we hone over time.

Our hunter-gatherer ancestors relied on their immediate impulses to survive—whether evading a snarling beast or spotting wild game. Anthropologists have suggested that perhaps we've inherited the penchant for seizing a moment's opportunity without regard for later consequences. Impulses still save our lives when we're fleeing a burning building or dodging a speeding car, but they fail us when we're trying to build a nest egg or get our neighborhood ready for the next wildfire.

Contemporary psychologists tell us that reckless decisions arise from people's default to a reflexive mode of thinking, known as "system 1." The rational and circumspect "system 2" mode of thinking is more taxing on the brain, and therefore more rarely engaged, contends Daniel

Kahneman, Nobel Prize–winning cognitive scientist. Neuroscientists point to the body's powerful limbic system, which governs our responses to emotions like fear, as the mechanism by which our immediate urges override caution about the future.

Plato wrote in his dialogue *Protagoras* in 380 BCE that it is the miscalculation of future pleasure and pain that leads to folly. In 1920, British economist Arthur Cecil Pigou similarly described humans' skewed view of what happens over time, calling it our "defective telescope." Today's economists call this pattern of making decisions "hyperbolic discounting," or present bias. The research of Harvard psychologist Daniel Gilbert shows that people fixate on what lies ahead but misgauge it. He argues that our view of the future is warped in part because we overestimate the impact of singular future events—like a job promotion—on happiness and underestimate the impact of accumulated minor events.

Each of these thinkers has helped the world understand why people make reckless decisions. But what they have told us is insufficient for the task of correcting course. In the game of telephone between such experts and the broader public, a misconception has emerged that recklessness is a fixed trait of human nature. That belief ignores the role of culture, organizations, and society—and the body of recent research findings that show we can influence and even quell recklessness. Humanity is more than mere biological programming: We can change our behavior through conscious decisions—from what we do in line at the grocery store to how we make laws. What appears our doomed fate is, in reality, a choice.

The dominant culture of our society fights against our aspirations to think ahead. We have come to expect instant gratification, immediate profits, and quick fixes to every problem. The battle cry of today's entrepreneur is to reduce the friction between craving and fulfillment, a cause that attracts ample money and talent: Search engines anticipate our queries before we finish typing them. As I tap my foot impatiently on the train platform, an advertisement on the wall from the ride company Uber boasts: "Good Things Come to Those Who Refuse to Wait." Urgency and convenience are dictators of decisions large and small.

In our era, we are frequently distracted from our future selves and the future of our society by what we need to accomplish now. Our in-boxes and texting threads fill with messages demanding immediate response. We measure ourselves and others by what we accomplish right away, whether it's making a sales target, winning a game, or scoring well on a test. As we absorb news flashes by the minute and an endless onslaught of social media updates from friends, our attention is focused on the incremental. We lose sight of what we want to do over time and ignore the future because we can—at least for now.

This need not be the case. The tradition of people coming together—in neighborhoods, communities, organizations, and countries—stems from our need for groups to help us do what we cannot easily do alone. Collectively we trade goods and skills, discourage violence and punish crime, educate the young, regulate currency, and feed the hungry. Cultural norms and institutional rules encourage people to act in their own best interest, and on behalf of the common good. Similarly, individuals could make better decisions about the future if organizations, communities, and society were designed to help us heed warning signs and weigh delayed consequences.

The problem is that today, such collective institutions are actually making it harder, instead of easier, for us to think ahead. The conditions we have created in our culture, businesses, and communities work against foresight. What could be bulwarks against myopia are instead holes in the hull, and we are flooded with immediate concerns.

Workplaces and schools reward quick results in the form of quarterly profits and test scores rather than what happens over time. The stock market and the popular vote favor ephemeral wins over down payments on future growth. Rapid replacement is codified into law—election cycles, for example—and engineered by consumer product designers in the form of gadgets that fast become obsolete. Governments rebuild in the same place after disasters, instead of preparing residents in advance. And policy makers chronically disregard future generations in their decisions.

The author Steven Johnson has posited that over history, people have gotten better at making long-term decisions. I disagree. The empirical evidence from our era—from corporate share buybacks to failed climate action—shows we are getting worse relative to the challenges we face, despite isolated exceptions. Johnson sees progress in prediction as evidence that we are planning ahead more than our predecessors; as we'll learn in chapter 1, prediction and foresight are far from the same. Where I do agree with Johnson is that we have tools at our disposal. The problem is that they are not yet widespread, and there is far more change required of us as a society and a culture than he suggests.

Instead of applauding our progress, we'd be wise to view ourselves as living in a society that rhymes with civilizations of the past—albeit with greater stakes and at grander scale. The geographer and author Jared Diamond studied civilizations that, after reaching their peak power and influence, rapidly collapsed. What those ruined societies across history—from the Polynesian inhabitants of Easter Island to Greenland's Viking colonists to the ancient Puebloans of the American Southwest—had in common, he argued, was their failure to heed warning signs of future consequences until it was too late. Poor choices—for example, to denude the landscape of trees or to avoid cultural exchange with groups with diverse resources—have doomed societies across history. When population growth and technological progress outpaced foresight, even the greatest of civilizations fell.

People alive today are not powerless when it comes to averting this kind of catastrophe. Even though we never know exactly what the future will bring, we can make choices to avoid the fate of failed civilizations. We can take steps in our own lives, as well as in businesses, communities, and societies, to get ready for the future and live with fewer regrets. We can forge new cultural norms, design better environments, and reinvigorate our best institutional practices.

In the coming pages, I will unearth the misconceptions that are holding us back from this task, and point to what we can do differently. The inquiry starts with the individual and family (part 1), then turns to businesses and organizations (part 2), and finally to communities and society (part 3). Each chapter offers a strategy for reclaiming foresight, and embedded within each chapter are techniques that can be used to achieve it.

Along the way, I'll examine how we privilege information-gathering over judgment in many decisions. I'll show how people, organizations, communities, and societies make reckless decisions even with good predictions on hand, and how measuring immediate results comes at the expense of perspective and patience. I'll uncover the folly of leaning too much on history—and the folly of ignoring it. I'll explore ways to overcome the paralysis born of doomsday prophecies, and to correct the myopia bred by modern business and political practices.

What you'll find if you read on is that some of the most important insights about acting for the future come from unusual places and overlooked fonts of expertise. The lessons in this book hail from a collegial team of women doctors, a study of Cameroonian toddlers, the exile of an ancient Athenian statesman, the redemption of a fallen financial hero, the survival strategy of an eighty-thousand-year-old tree colony, the rebuilding of a Shinto shrine, and the lingo of professional gamblers. There is much to be learned from the work of the very young—a seventeen-year-old Eagle Scout, and teenage plaintiffs suing the U.S. government—and from the very old—octogenarian farmers and long-dead poets. The rock stars of thinking ahead are not always whom you might expect.

PART ONE

The Individual and the Family

GHOSTS OF THE PAST AND FUTURE

Imagination as Power

In this world
we walk on the roof of hell
gazing at flowers

—KOBAYASHI ISSA

Nothing makes the mind flash ahead quite like a walk through a graveyard. The sight of a tombstone is a stark reminder that, someday, the jig will be up.

On a recent visit to the sprawling Mount Auburn Cemetery in Cambridge, Massachusetts, I encountered the tombs of the poets Henry Wadsworth Longfellow and Robert Creeley, and of the celebrated inventor Buckminster Fuller. Centuries-old beech trees towered overhead, and wild turkeys strolled unaware over the remains of literary giants.

Most striking to me, however, were the simple tombstones marked with a name and a year of birth, followed by an open-ended dash. Any grave, like a fatal car crash on the highway, invites you to ponder your own demise. A stone that bears your own name must be more of a demand than an invitation.

Rarely do we stand before anything that confronts us with the many points of our lives to come *before* our deaths—what's ahead of us, but not at the very end. We don't often come face-to-face with our living future.

When we drive, however, we find a useful metaphor. The road as the dash mark between birth and death, lifted from the etching on the gravestone. The road ahead is the future.

What would it take to look farther down the road more often—to envision what lies ahead? And could we do so not with dread but a sense of a choice in the matter?

To exercise foresight on an actual road trip, we need more than a map and a GPS, more than road signs alerting us to blind curves, beautiful turnoffs, and falling rocks.

Say this is a road with hairpin turns, carved into a cliffside, overlooking the ocean. We face choices before we take this trip—whether to go by day or by night, whether to wait for fair weather or drive in a hailstorm, whether to fill the tank and check the tires, whether to take a friend. What each of us decides should depend on what we value and how we tolerate risk. We'll need to gather data—road conditions, gas stations, points of interest on the route. But no matter how much information we amass, we won't know everything about what is ahead. A drunk driver, a wayward child crossing the road, a flock of migrating birds overhead, a vista that calls forth a childhood memory.

There are endless possibilities in the future, as there are on the road we have yet to travel. Without perfect knowledge, we make decisions—consciously or unconsciously—before we depart. For the trip of our lives, it is not wise to just stare at the speed gauge on the dashboard or at the pavement in front of us, and it is not quite enough to look at signs or numbly follow the voice directions for each turn. We need to imagine what we'll face and act on behalf of our future selves. We need to envision the road and the way we hope to navigate it.

The commander of the Roman naval fleet had just finished eating lunch when his sister spotted a strange black cloud swelling above an eastern mountain. Known as Pliny the Elder, the military leader lived by a doctrine of curiosity. In his spare time he named plants, birds, and insects—an ancient forerunner of modern naturalists. As the mountain's ink spilled into the sky, he summoned his ships so he could get a closer look.

As Pliny set out from Misenum on that day in the year 79 CE, however, a messenger intercepted him. *Come help us,* urged a fearful resident of Pompeii, who saw the cauldron of smoke brewing above Mount Vesuvius.

The volcano soon spewed plumes of flame, its ridges cascading in a flood of ash and stone. The sky rained pumice and cinder, and the air reeked of sulfur. Homes shook until they collapsed. Roads crumpled like paper meant to stoke a fire. People fled with pillows tied to their heads by dinner napkins, a poor defense for the crush of rocks.

From what layers of ash have preserved, we know that bacchanalia raged along the Bay of Naples until Vesuvius broke up the party. Archaeologists have found broken dinner plates, brothels, and the bones of gladiators and children in the ruins. Phalluses, as if useless talismans, appeared everywhere—painted into frescoes, carved in roads, sculpted in mosaic.

Pliny the Elder died trying to rescue panicked residents of Pompeii. His nephew, a budding poet who chronicled the eruption, escaped by coincidence, not foresight: He chose to finish his homework instead of accompanying his uncle.

The horrific eruption of 79 CE was not ancient Pompeii's first catastrophe, however. Often overlooked in popular accounts is that a severe earthquake had hit the Italian region of Campania seventeen years earlier. The quake flattened buildings and killed a flock of a hundred sheep. Statues of the gods cracked. Yet by the time the earth trembled in the weeks before the Vesuvius eruption, the foreboding of tremors had been forgotten, and few in Pompeii suspected imminent danger. Or at least that was the account of Pliny's nephew in letters he penned to Tacitus a quarter century later.

In ancient Rome, natural disasters were widely seen as the work of angry gods—a belief that the era's philosophers sought to dispel. Seneca was perhaps the most admired thinker of the age, also a tutor and adviser to the brutal emperor Nero. In the year 62 CE, shortly after the quake hit Pompeii and nearby Herculaneum, Seneca wrote about earthquakes in *Natural Questions*. Earthquakes, he concluded, were caused by air violently freeing itself from subterranean caves by breaking through rock. Pliny the Elder concurred: Earthquakes were unquestionably caused by winds.

"All places have the same conditions and if they have not yet had an earthquake, they none the less can have quakes," Seneca wrote after the disaster, in an effort to calm the fears of residents. "Let us stop listening to those who have renounced Campania and who have emigrated after this catastrophe and say they will never visit the district again."

Stay put, Seneca advised. And don't worry about anticipating where or when an earthquake could strike—it's just too unpredictable.

We can forgive Seneca for the blasé attitude—based on available knowledge at the time—just as we can excuse the thousands of people who perished in Pompeii when Vesuvius erupted a few years later. They didn't see it coming.

In their era, even the most logical explanations for natural disasters resembled the late-night musings of stoned old men more than scientific hypotheses. Pompeii's ancient residents lacked the knowledge we have in retrospect: The mountain that towered over the region was a volcano. Previous quakes signaled a dangerous seismic zone, and the tremors warned of a coming eruption.

But what if the wise men and women of Pompeii *had* known that Vesuvius was about to blow? Imagine if, instead of reassurance, the philosopher Seneca had warned that the beautiful Bay of Naples was too dangerous to inhabit. Or suppose that the commander, Pliny the Elder, had urged precaution and created an evacuation plan, instead of being caught by surprise. Would they have been able to avert the catastrophe?

We will never know the answer to such questions. But we do know that knowledge about dangers ahead, even today, is often not enough to persuade people to act on behalf of their future selves or their families.

People have long sought to predict the future. The quest to know what lies on the road ahead is as old as recorded human history, dating back to the early trackers of celestial patterns that marked changing seasons. Methods have ranged from the empirical—observing stars in the night sky or the position of the sun rising on the horizon—to the mystical—oracle bone readings and riddles delivered by soothsayers. The ancient Romans routinely searched for signs of coming calamity: hens that refused to eat, sweating statues of deities, and fiery night skies that today's astronomers call the aurora borealis. Similar divinations persist in our time, in the form of octopuses believed to foretell World Cup match outcomes, carnival palm readers, fortune cookies, and Magic 8-Balls.

We still view the future with limited vision, just like the ancients. But people have never lived in a better era for forecasting natural hazards—even if election and sports predictions still fail us too often. We may not always know the precise day or week that an earthquake or flood will strike. But we do know far more about the future dangers in a given place than previous generations did.

Our era boasts knowledge of plate tectonics, the shifting crust of the earth, and the records of past eruptions and storms. We have instruments that monitor seismic activity, and early warning systems for tsunamis that get improved upon after each disaster. We don't have to be fatalistic like Seneca, who deemed every place equally dangerous.

Just in the past fifty years, the oft-bemoaned weather forecast has become dramatically more accurate—giving a heads-up to baseball fans and those of us who suffer bad hair days. In the nineteenth century, newspapers unfavorably compared weather service predictions to the prophecies of astrologers; the latter were often more reliable. Today, people like me cling to forecasts as if they were the gospel delivered directly from God with hourly precision to their smartphones.

We owe our reliable forecasting tools today to the Enlightenment and its aftermath: ideas tested by generations of experimenters, data collected over vast expanses of geography and time, instruments that observe nature in granular detail, and knowledge that sprawls endlessly, its tendrils reaching to the far corners of the world and back again. We tend to look back on the victims of Pompeii with pity, armed with the superior knowledge and sophisticated technology of our time.

In the twenty-first century, we've entered a new era of obsession with prediction. This is in part due to a revolution in data analytics, enabled by leaps in computing power that have outpaced engineers' wildest expectations, and by the trillions of data points collected by billions of instruments around the globe—like an army of celestial trackers now observing virtually everything on Earth and in near space.

Our appetite for prediction is being served by advances in machine learning, the ability of computers to learn from past patterns to project future trends. People now use such tools to predict what books or clothes a given customer will buy, how a flu outbreak will spread, which city blocks will see crime spikes during power outages, and how an oil spill will disperse at sea. The focus of prediction in our era is often the immediate future—what song will be the next hit or what ad will capture your attention when you visit a certain website. The latest trend in weather forecasting apps is to try to predict more accurately whether it will rain in the next hour, not whether it will be sunny in ten days or two weeks.

Even these scientific predictions are inherently imperfect because they rely on past trends to foretell the unknown future. Yet they are useful enough to attract billions of dollars in investment and to seize the attention of corporations and governments as the bandwagon of the decade to jump on.

Robust predictions of the future are worthwhile, of course, only if we use them. The paradox is that people are often not swayed by them to prepare for the future—that is, to move out of the proverbial volcano's path. A good forecast, it turns out, is not the same as good foresight.

In my experience, it is easier to contemplate death by shark attack than it is to envision myself with fake teeth.

The thought of aging brings up a feeling of dread—a cue to check my e-mail or rearrange a cluttered closet. Being old is not something that I've experienced yet—and the episodes in my past are thus far limited in helping me conjure the future. Yet I know that the odds are, given my gender and family history, I'll be getting old someday. Very old. I have on hand a fairly reliable prediction based on robust historical data. What I don't have is the foresight to use it.

I've struggled for years to make decisions that will shape my experience of old age—whether or not to have kids, how to save, how to invest. Some might say I've actively avoided making certain pivotal choices on behalf of my future self. I'm not alone in this predicament. A 2006 survey of hundreds of people from twenty-four countries found that most people do not think past fifteen years into the future—even though they worry about the future more than the present. The future "goes dark" in their imagination at some point between fifteen and twenty years from the present moment.

Humans have roamed the Earth for more than seven thousand generations, and for much of that history, they could expect to die around age forty. Over the past two centuries, however, the average life span has grown steadily with medical discoveries that have kept infants from dying and cured people of infectious diseases. Today, life expectancy at birth is on average more than seventy years worldwide, and above eighty years in most of the developed world. By 2100, newborn babies in the developed world may well expect to live a century or longer.

This is mostly good news for humanity. Our minds, however, have not yet evolved to easily contemplate our long life spans. That's why it's rare for people today, especially if they are relatively young, to envision the concrete reality of getting older.

For a society that is aging, with growing needs for caregivers, savings accounts, and retirement plans, the failure to imagine old age poses a considerable threat. How will generations alive today ever plan for the long years ahead if we can't even fathom ourselves in the future? In countries including the United States, Canada, Germany, and Japan, personal savings rates declined in recent decades, even as life expectancies rose. A fleeting glance at the U.S. lottery, funded to the tune of billions of dollars mostly by people who earn very little and have little money to spare, belies the idea that people simply save less because they earn less. In buying lottery tickets, people entertain a fantasy of winning large sums of money in the immediate future, a hallucination fed by news coverage of jackpot winners and by small payouts earned from scratch cards. By contrast, we don't tend to daydream of our likely long-run future of fighting back pain while playing with our grandkids.

A few years ago, an economist named Hal Hershfield, now a professor at UCLA, became interested in getting young people to save for their future. He had an idea: What if he could create an imaginary, sensory experience of old age? By collaborating with a virtual reality designer, Hershfield created a program that generated avatars of each student in the experiment as an elderly person—with wrinkles, receding hairlines, and gray hair. To look at the avatar in the virtual reality room was to look in a time-traveling mirror, in which each of your gestures was mimicked back to you by an older version of yourself. The goal was to try to get the college students to imagine the experience of their future selves.

At Hershfield's suggestion, I gave a more rudimentary technology a try, a smartphone app called Aging Booth that manipulates a photograph to age the person pictured. I wanted to see what my own reaction would be to seeing myself as an old lady—if not as an active, gesturing avatar, then at least as a frozen image. I considered taking a selfie headshot in my office, but then thought better of being at close range. Instead, I fed the app a flattering mid-length photo of me standing in front of a water-fall in Oregon.

The doctored image that emerged somehow both horrified and comforted me. I recalled the moment in a recent *Star Trek* movie when Spock encounters the older version of himself, who has been roaming the universe for hundreds of years. The old Spock imparts crucial knowledge to the young version. As I looked at my face in the aged photograph, the eyes sunken into black saucers, deep contour lines creasing my forehead and jaw, I saw someone both familiar and alien. I wondered what wisdom that person would have that I had yet to find. I embodied her, just briefly. It felt like me, and I even hoped it would be me, if I were lucky enough to ripen to eight decades. My imagination was activated, and so was my empathy for my future self. The imaginary future suddenly seemed more real, because I felt it in a personal way and saw it manifest on my own face.

In his experiments, Hershfield has found that college students who come face-to-face with themselves as elderly people are afterward more willing to save money for retirement than those who do not encounter their virtual reality avatars. They are more willing to save than students who merely encounter photographs of other old people, and to save more money for retirement than those in control groups who already save money.

Hershfield's research is among the new efforts to help people better imagine what they have not yet experienced—or what they no longer remember. Virtual reality is being deployed to give wealthy people a brief simulation of being homeless, to invite Alzheimer's patients to walk through their childhood homes, and to prepare professional athletes for encounters against particular opponents on the football field or basketball court. The author and therapist Merle Bombardieri uses a low-tech version of these exercises with couples deciding whether or not to have children. She asks them to imagine themselves at age seventy-five in rocking chairs, and to play out different scenarios to see which elicits the least regret.

What I find intriguing about such tools is that they do not merely focus on conveying facts about the past, present, or future. They try to mimic the experience of a different time or context. My encounter with an aged photo actually made it harder to ignore my future self—and it was only later that I understood why.

The way the students in Hal Hershfield's experiments responded to their virtual reality avatars stands in sharp contrast to how people today respond to hurricane forecasts.

Tropical storms could be the poster children for our era's scientific prowess in predicting disasters. When Christopher Columbus was exploring the West Indies in the fifteenth century, and even in the eighteenth century when British seafarers were building an empire, hurricanes struck ship captains by surprise, wiping out entire convoys. In 1900, before the deadliest hurricane in U.S. history destroyed Galveston, Texas, the leading National Weather Service forecasters did not even issue the city's residents a warning.

Today, meteorologists can map out the hundred-mile radius of a hurricane's landfall, and can often chart the likely trajectory of a tropical storm early enough to give us seventy-two hours' notice to flee or hunker down. We know days in advance whether we could be hit.

How people react to severe hurricane forecasts, however, exposes the limits of this progress. Wharton economists Howard Kunreuther and Robert Meyer have found that most people, when faced with forecasts of deadly storms—such as Hurricane Sandy in the New York region and Hurricane Katrina on the Gulf Coast—do little more than buy bottled water. Nor do they make advance preparations to elevate their homes, seal foundations, or waterproof their walls, despite living near the ocean. The majority of people who live in regions struck by hurricanes do not buy adequate flood insurance. Good information about the future has not ensured good judgment.

Damage wrought by hurricanes is on a steep rise, with costs in the hundreds of billions of dollars in the United States alone in the past decade. The global cost of hurricane damage is expected to rise from

$26 billion per year to $109 billion per year by the end of the century. One reason for such skyrocketing costs is that people continue to build homes and businesses in the likely path of storms. Another reason is that people who live on coasts do not sufficiently prepare for severe hurricanes, even when they are fast approaching.

Our failure to prepare is not limited to hurricanes. In the United States, only 10 percent of earthquake-prone and flood-prone households have taken actions that would reduce their property losses from disaster. The problem is not just American, but global. From 1960 to 2011, more than 60 percent of the damages from major natural catastrophes around the world were not insured. It is not just a matter of rich or poor, either: Only half of the damages from earthquakes, tsunamis, and floods were insured in high-income countries over that same time period.

The choices we make today shape tomorrow's so-called natural disasters. Dennis Mileti, a disaster expert who formerly directed the Natural Hazards Center at the University of Colorado, Boulder, believes *people* create what are often deemed "acts of God." In his view, nature supplies the threat of an earthquake or hurricane, while true catastrophes are owed to poor human decisions about where to settle and how to prepare. "Leave God out of it," he quipped to me when we met over cocktails in 2017. "God had nothing to do with it."

It's a common misconception that people fail to get ready for future threats because of a lack of knowledge. A survivor of Hurricane Katrina who was trapped on the roof of his coastal Mississippi home during the storm helped me realize that the problem is often not a matter of awareness—or even a matter of resources. Jay Segarra was the chief of pulmonary medicine at the Keesler Air Force Base hospital in Biloxi, a well-respected doctor also versed in the science of ocean currents. Yet he didn't take warnings that promised the worst hurricane in the area's history all that seriously. He believed he could weather the storm with a generator and flashlights, just as he had those in the past. He didn't have flood insurance. And after the historic 2005 storm, Segarra rebuilt his home in the exact same spot—about a football field's length from the Gulf of Mexico.

Segarra finds the idea of educating people in harm's way useless. "People on the coast know that the government—the taxpayers—are going to make them whole," he told me, speaking of the relief funding that pours into privileged communities after disasters. If he and his neighbors had been required to buy flood insurance at full cost or to cover their own losses from storms—or had they been prohibited from living so close to the coast—it may have been a different story, he said. "Then and only then will people stop building in vulnerable areas."

Government programs discourage people from preparing for the future—bailing us out after disasters and helping us rebuild rather than encouraging relocation. And most warnings of disaster are not paired with enough resources to help the poor buy supplies or evacuate. Reckless decisions thus can't be entirely pinned on individual people and families. American culture, moreover, breeds unbridled optimism about our future fate: We feel destined for brighter days while failing to gaze through the telescopes that would let us see the storms on the horizon. (I will turn to the ways that communities and societies can change such practices and norms in part 3.) There is still more that we could do as individuals, however, to get ready for disasters, if we had the foresight.

Jay Segarra had no reason to prepare—unless he could have imagined that during Hurricane Katrina, he was going to spend a harrowing day clutching the skylight on top of his house in hundred-mile-per-hour winds, or that floodwaters would carry away family photo albums and a beloved heirloom cello built in Paris in 1890, once played by his father. If he had been able to fathom that, he said, he would have evacuated.

The conundrum for Segarra was that what was foreseeable in the future—and even what was precisely predictable—was not conceivable.

Whenever I tried to convince business executives that they should prepare for droughts and heat waves, I armed myself with reliable projections of the future. But corporate leaders, like Jay Segarra, struggled to see themselves and their companies in the forecast scenarios. In the short run, of course, the companies had plenty of profits on the line. Yet it was still mysterious that their leaders did not grow concerned about severe and costly future threats. The science of risk perception helps explain their complacency.

People often take in information that aligns with what they want to hear about the future and filter out what does not. We tend to overestimate how long we will live, how successful we will be, how long our marriages will last. It's a kind of denial of reality. When a predicted disaster turns out to be a near miss or less catastrophic than expected, it reinforces our sense that nothing bad will happen to us, even if the past is a poor precedent. We come to view forecasters as boys crying wolf.

Blind optimism of this sort can paralyze us when it comes to planning for future danger. During the 2012 hurricane season, for example, Bob Meyer and his colleagues studied in real time how residents of Louisiana and New York responded to predictions made for Hurricane Isaac and Hurricane Sandy, respectively. They found that people misgauged the potential damage to their property and their homes, despite robust warnings about the threats posed by wind and storm surge—and even prior knowledge of living in high-risk areas for flooding. People also underestimated the time that they would endure power outages, by days. Relatively few made evacuation plans, purchased power generators, or put up storm shutters—even among those who already owned the shutters. Small fractions had made previous home improvements to withstand flood and wind. Nearly half were underinsured. Few, particularly in New York City, had even contemplated the possibility that their cars might be submerged in floodwaters. This was despite extensive media coverage that warned of life-threatening floods—and record-breaking viewership of forecasts as the storms made landfall.

It's human nature to rely on mental shortcuts and gut feelings—more than gauges of the odds—to make decisions. That was probably as true for ancient Romans as it is for us today. These patterns of thinking, I have learned, explain why all the investment in better prediction can fall short of driving decisions about the future. Even if Pliny the Elder and

Seneca had today's forecasting tools, they might not have prevented the destruction of Pompeii.

The threats that people take most seriously turn out to be those we can most vividly imagine. At airports in the 1950s and 1960s, insurance underwriters sold accidental death policies to people just before they boarded planes. Companies made huge profits selling such on-the-spot plans—when the risk of dying in a plane crash seemed imminent and imaginable to people. Swiss economists Helga Fehr-Duda and Ernst Fehr contrast that phenomenon with how people around the world have failed to buy insurance for natural catastrophes over the past fifty years.

I once heard the acclaimed filmmaker Wim Wenders describe what gets people's attention: There is, he said, "a monopoly of the visible." Behavioral scientists such as Daniel Kahneman call this skewed pattern of human perception the "availability bias" and note that it contributes to people misgauging future risks. Kahneman points out how the fear of unlikely events, such as terrorist attacks, or unrealistic hope for future opportunities, such as winning the lottery, are reinforced by the ease with which we can summon images of those future scenarios.

When a speculative scenario of the future has more colorful details, it seems more likely to us, whether or not it actually is. When sensory details do not surface, on the other hand, it can make a future scenario seem unlikely or even impossible. Kahneman discovered that if you ask people to compare the chances of an earthquake in California happening sometime next year with those of a catastrophic flood somewhere in North America happening in the same time frame, they will incorrectly pick the former as more likely, just because it's more geographically specific.

Movies about terrorist plots and media coverage of lottery winners can turn our attention to what is rare. A shark attack can seem more likely than the common hazard of slipping in the bathtub or the likely realities of aging. We believe we can win the Powerball, and we underestimate unfathomable sea-level rise.

When it comes to natural disasters, we might expect intense news coverage of their aftermath to sway people to prepare more. And immediately after storms and earthquakes, people in affected areas do tend to buy more hazard insurance. But media blitzes after disasters happen in brief stints, then fade away. The events often lack familiarity, as if happening to distant people in distant places, and not to us. As years go by without incident, people in former disaster zones stop buying insurance—just when they are likely to need it the most.

Forecasts of coming storms do not usually remind us of the concrete consequences of failing to prepare. It might help if, instead of just tracking incoming storms, broadcasters showed us images of past devastation in our communities.

Prediction is not that helpful for heeding future threats, unless it is paired with imagination. And if we can't incite our imaginations to serve our purpose, today's revolution in scientific forecasting might be in vain. When it comes to threats we cannot predict, moreover, we will be at a complete loss to plan ahead if we do not expand our view of the possible.

When people imagine the future, as I did when staring at the image of my aged future self, we rely on abilities common to all human beings. Foresight is therefore possible to decode, and attain. Scientists have intriguing insights about why people are able to envision the road ahead when we do.

That humans contemplate the future at all is somewhat miraculous. Most animals seem to indulge in what they can, whenever they can get it, without entertaining the idea of future consequences. Since we can't perceive what lies ahead with our senses, it takes an imaginative leap.

Some evolutionary psychologists believe that the ability to imagine the future may actually be what distinguishes being human, and what has made possible our dominance over faster, stronger beasts in the animal kingdom. Thomas Suddendorf, a University of Queensland professor, has researched the origins of human prospection—the capacity to think creatively about future possibilities. He believes that what sets humans apart is our ability to concoct scenarios that have not yet happened and embed ourselves inside them—a feat in light of the competing temptations that we encounter continuously in the present. Simulating an episode in the future motivates us to plot battle strategies now that will help us prevail over adversaries later. Over history, we have had to outsmart other animals because we could not always outrun or overpower them.

In recent years, several researchers, including Suddendorf, have advanced the idea that this ability to anticipate what lies ahead relies—at least in part—on our memory: our power to rearrange episodes of the past in our imagination.

To understand how this works, imagine yourself in a moment you're looking forward to in the future. Maybe it's walking your daughter down the aisle at her wedding, or the handoff of your degree on graduation day. Perhaps it's the first time you will scuba dive, or step foot in the Roman Forum. You can likely summon the images, like a movie projected on the screen of your mind. You may even faintly hear the birds chirping or the crowd applauding.

What you've just exercised is the human habit that cognitive scientists call mental time travel: using memories to propel the mind into future moments. When you cast your thoughts ahead like this, you are refashioning images and sensory details of past episodes—what you've experienced in childhood, seen in movies and photos, and heard about in others' stories. You didn't have to experience those exact moments in the past to be able to see yourself in the future ones.

In the fantastical world beyond a mirror that Lewis Carroll portrays in *Through the Looking-Glass*, the White Queen tells Alice that she remembers only the future—what will happen the week after next. In a sense, remembering the future is what we all do. Thinking ahead relies on episodic memory—our ability to recall scenes, not just remember facts or skills—to anticipate what we have not yet lived. We can do this even though we warp the past like unreliable eyewitnesses to our own triumphs and defeats.

Severe amnesiacs typically cannot construct episodes of either the past or the future, Daniel Schacter, a professor of psychology at Harvard, told me. They come up blank if asked to recall a friend's wedding from the past year, or if asked to imagine attending a wedding next week. In Schacter's view, this shows that humans likely evolved to remember past scenes for the survival skill of imagining what possible threats or opportunities lie ahead. This function of memory could explain why we have such imperfect recall—the kind that leads people witnessing a crime to incriminate the wrong person and gets us into fights with our spouses about what exactly happened at that dinner party seven years ago. It's the gist that matters when it comes to repurposing the past for the future, not the exact details.

Mental time travel is aided by certain habits. Wellesley College psychology professor Tracy Gleason believes that one is letting the mind wander. It may be that the wandering mind has more latitude to harvest and reassemble episodes of the past.

Gleason studies imagination in children. (Many famous writers, she tells me, had imaginary friends in childhood.) Over coffee one day in 2016, she described to me a camping trip in Colorado she was planning with her family. They are not avid campers—in fact, her husband dreaded the idea of sleeping in the great outdoors. To prepare for the trip, she walked through a day in her mind, imagining a challenge her family might face or adventure they might experience. *How will they drink coffee? Time to pack a portable coffeemaker. What will the kids do on the road trip to the campsite? Need to pack some games for the backseat. What if they encounter a bear?* She had to both conjure and plan to pull this off. Rather than predicting the future with algorithms like a modern forecaster, Gleason was engaging in a free-form process of scenario generation that was only as good as how far she let her mind wander. The goal was to creatively anticipate future events, including dangers and opportunities.

Gleason acknowledges that for some people, imagining the future like this can induce paralyzing anxiety—making them worry about all that could go wrong. The key, she says, is to try to see yourself as having the ability to solve the problems that come up in the future. In other words, one way to make your dread productive is to imagine yourself reacting to what happens in the future with success.

This suggests it might help people to imagine not only what could happen to us during a coming earthquake or storm, but also how we could handle whatever we face—with actions we can take ahead of time and in the future moment. While predictions of disaster might make us feel like victims, using foresight to act on those predictions can make us feel like the heroes of our own unfolding stories.

When our minds wander, we are thinking about something unrelated to the demands of a given moment. Psychologist Benjamin Baird, who studies what people think about when daydreaming, has found that we have spontaneous thoughts predominantly about the future—to help us plan for future situations. On the other hand, giving people tasks that place high demands on our cognition, requiring deep mental focus in the present, tends to constrain the amount of future thinking we can do. Mind-wandering can derail us from urgent work but offers the gift of making prospection possible.

When we successfully imagine the future, it can come alive in our senses in the present. It can even motivate our current choices. Over the past decade, studies have found that when you invite people to generate detailed scenes of specific future events, it can combat impulse buying of booze among alcoholics, encourage teenagers to have more patience, discourage junk food snacking, and lead to healthier, lower-calorie food choices among obese and overweight women in food courts. Seeing future events in lurid or lovely detail can sway us today because of how it makes us feel.

The imaginary future can also drive us to persevere, enduring pain now for the sake of what we can get later. Picturing graduation day might help you, for example, get through studying for exams and writing final papers. This power also has nefarious potential: Dictators have in the past ably exploited the envisaged future—an afterlife where one is liberated from toils, for example—to keep people placated amid present suffering.

Social movements have drawn on the motivational power of imagination, too. In 2017, I went to see Marshall Ganz, a sociology professor at Harvard's Kennedy School of Government who had been my adviser in graduate school and is something of a guru to thousands of activists around the world. He studies the inner machinery of movements for social justice.

Ganz believes that imagination of the future is critical to the success of social movements because it inspires people to keep going amid setbacks. And he has experienced it firsthand.

Ganz was raised in Southern California, the son of a rabbi and a schoolteacher. As a young college student he went to Mississippi in 1964 to register black people to vote during what's now known as Freedom Summer. He had no interest in going back to school when he saw the work to be done helping people secure basic human rights, and so he dropped out to become an activist. Soon he joined the United Farm Workers movement in California, serving as a key strategist for Cesar Chavez. Only twenty-eight years later did he return to his formal education, finishing his undergraduate degree and then a PhD after decades of hands-on experience with pivotal social movements of the twentieth century.

"The vision of the future has to be concrete enough that you can visualize it," he said. "In the farm workers movement, it was access to toilets in the fields for the workers, not having to pay bribes, and having medical care. It was not abstract." The most successful campaigns of the U.S. civil rights movement, he says, similarly motivated activists with specific imagined futures—blacks sitting at the same lunch counters as whites or riding where they pleased on the bus. The scenes were not predicted events on particular dates, but rather animating visions that motivated people to endure the short-term physical hardship and the violence they experienced during boycotts and demonstrations.

Martin Luther King, Jr., used to watch the original televised *Star Trek* series with his family, in part because it portrayed an imaginary future in which a black woman, Lieutenant Uhura, was the fourth in command on the starship. At an NAACP fundraiser, he once told the show's leading actress, Nichelle Nichols, that her portrayal of a brilliant black woman as an equal to men of other races was breathing life into the civil rights movement, making the protesters' sacrifices in the present seem destined to yield progress toward equality. A vision of the Promised Land, articulated in religious scripture, also bolstered the movement's imagination of a different future.

We know that an imaginary picture of the future can be motivating even to people who may never live to see the fruits of their labor. Countless civil rights activists did not survive to elect the first black president of the United States. Similarly, a group of retired NASA scientists in their eighties have been working to build machines to help future humans breathe on Mars, because of feats they imagine beyond their lifetimes. When I first heard about these scientists, I was in awe, since I had yet to even envision the span of my own life. The road ahead seemed darker to me than their view of outer space.

The human talent for simulating episodes in the future, while impressive relative to our wild beast brethren, is rarely engaged and limited compared with the demands of our world. When constantly focused on what's right in front of us, as is common in our era, it becomes difficult to let our minds wander and to concoct future scenes.

We do not fully inhabit futures that are unlike the past, and past experiences—much like the earthquake that preceded the eruption of Mount Vesuvius—are not always enough to conjure the future. Our minds can wander only so far.

In Charles Dickens's enduring novella *A Christmas Carol*, it takes a visit from the Ghost of Christmas Yet To Come for Ebenezer Scrooge to become fully horrified about the consequences of his current behavior. He needs help inhabiting both the past and future in his imagination—not to mention the present beyond the walls of his home. His visits with the three ghosts change him, and he becomes willing for the first time to share his wealth with people in need.

To boost our ability to imagine the future, we need to find our proverbial ghosts: We need practices and tools that help us inhabit the experiences we have not yet had. We need ways to gaze farther down the road.

After learning about Hal Hershfield's study using elderly virtual reality avatars, I sought to know more about technologies that could aid the imagination. At Stanford's Virtual Human Interaction Lab, researcher Jeremy Bailenson is creating simulated environments where people can fly like Superman over a city to feel powerful or altruistic, virtually eat the amount of coal needed to heat their daily showers to sense viscerally what it is to consume fossil fuels, and experience themselves in the body of someone of a different race.

Fit with a headset in a small square room at the VR lab in 2016, I was transported to an industrial warehouse. I stood on a wooden plank, no more than a foot wide, that towered more than thirty feet above a cavernous pit. I walked the plank. A researcher then asked if I'd like to jump off the plank into the pit. I hesitated. I've jumped off higher cliffs than this in real life, but into water.

I knew in a theoretical sense that the experience was not real, but my body was fooled. I trembled and felt my heart race, then finally went for it after mustering the full force of my will. I stumbled and "landed" as if I had actually jumped from a great height. I felt better about being so afraid when I later learned that a third of the thousands of people who come to the lab refuse to jump because of how real the experience feels.

Next, I swam in a coral reef that was vibrant with colorful fish, as if I were a scuba diver. The simulation flashed forward to a reef with dead coral and no fish—a model of coral reefs in the year 2100 if humans do not stop polluting the atmosphere with carbon dioxide, which warms and acidifies the ocean. Before I left the lab, I got to see myself as a buxom blonde wearing scant clothing, as an old white man, and as a black woman being yelled at by an older man in a business suit. All three of the avatars surprised me in ways I didn't expect. I felt they were more than intellectual experiments or gee-whiz moments with gadgetry. Getting yelled at as the black woman and gawked at as the blonde aroused my emotions—sadness, fear, frustration, pride, humiliation. Being the older white man made me feel powerful. I still of course can't say I know what it's truly like to be any of those people, but I got closer than I had been before.

One reason these experiences felt real, even though my rational mind told me they were simulated, is that sounds and vibrations reinforced the imagery I was seeing and the movements I made, something Bailenson and his colleagues call "haptic feedback." The floor shook when I jumped into the pit, because of speakers below my feet called "buttkickers" that delivered booms triggered by sensors on my wrists and ankles. Airplane-grade steel lined the floor, conducting vibrations the way pipes conduct water.

The technology convinced my brain I was not on a carpeted floor in a drab conference room in Palo Alto, but rather plunging into the dark depths of a warehouse pit or swimming in tropical waters. Bailenson has demonstrated in research studies that people who swim in the two coral reefs—one present and one in the future—in simulated reality sustain greater concern about coming threats to the world's oceans than people who simply read about the dangers of ocean acidification. Their concern also lingers longer than in those who watch films about environmental dangers. He believes this is because they have had an experience that affected them emotionally and physically, and even potentially imprinted on their memories.

It's unclear the extent to which such technologies can influence people's judgment and shape our decisions. But the tools are already being put to use commercially. Bran Ferren, the former president of R&D for Disney's Imagineering studio and the CEO of the technology and design company Applied Minds, created an astronaut-like aging suit. The suit blurs people's vision to simulate cataracts and macular degeneration and stiffens their joints to simulate arthritis—all so that they can experience sensations of old age. A company that sells elder care insurance uses the suit as a prop to encourage people to think about long-term risks they face in their lives—and to buy insurance. The Golden State Warriors recruited NBA star Kevin Durant in part by giving him a virtual reality experience of his potential future life in the San Francisco Bay Area. Several NFL teams now use virtual reality environments to let players carry out plays in a variety of scenarios against opposing teams, creating matchups and situations beyond what they can do on the practice field.

Virtual reality tools are also being used to train first responders for disasters like hurricanes and terrorist attacks, so they can practice how they might respond. Nurses and medical staff in Iraq have learned to triage patients with VR during battle, and EMTs have learned to respond to situations where there are mass casualties, like the Boston Marathon bombing.

Researchers have collected biometric data from people during high-quality simulations and found that they often react as people having a lived experience, their pulses racing, adrenaline surging, and blood pressure rising and falling. In this sense, virtual reality might do more than trigger imagination of future scenarios; it might have the potential to trick the body into *feeling* the future.

These technologies have yet to be used in a widespread way to help everyday people imagine the future risks of natural disasters. But with falling costs of VR headsets and the growing ease of creating simulations, it's possible we'll be able to use these tools as imagination aids to forecasts and warnings. Bailenson is working to make this happen. "In virtual reality, disaster is free," he says. "And nobody gets hurt." Cities might create simulations that help people realize that they can't move their cars through flooded streets or that their homes will be more damaged if they don't put up storm shutters. Communities could even help people envision hazards before they build or buy homes in fault zones or on beaches.

What interests me about such tools is not that they are a panacea, but that they show it is possible to design technology geared for future imagination, not just instant gratification. New tools illuminate our power to conjure the future—like lighting up the projector in the movie theater of our minds. Designer Anab Jain, the founder of a London-based futurist company, Superflux, describes a time she struggled to persuade government and business leaders in the United Arab Emirates to consider a scenario where they might want fewer cars on the roads of congested cities such as Dubai and Abu Dhabi.

"I cannot imagine that in the future people will stop driving cars and start using public transport," one man told her, looking at her models of the cities. "There's no way I can tell my own son to stop driving his car." Jain then presented the leaders with a vial of toxic-smelling air she made in a lab, a simulation of polluted city air in the year 2030. The next day, she said, the leaders made an announcement to invest in renewable energy.

Imagination of the future does not require the novelty of sophisticated technology. Without a VR headset or a chemistry lab, people can create focal points to envision the future, like calling up Scrooge's ghost.

Mere suggestion may be enough to invoke certain ghosts. For example, when Princeton University economist Elke Weber asked hundreds of people to describe how they wanted to be remembered by future generations, she found that they would subsequently make wiser decisions with respect to the future—including donating to climate change causes. Researchers in Germany similarly found that when people were asked to imagine concrete details in the daily life of a hypothetical young woman living on a hotter planet in the twenty-second century, they showed greater interest in climate change than those who merely learned about the woman's predicament from a slide show. You might call it provoking imaginative empathy.

Drawing on such research, two graduate students in Massachusetts started a project in 2015 to aid people's imaginations of the future. Trisha Shrum, an economics scholar, and Jill Kubit, a climate organizer—both mothers—were particularly concerned about climate change and its impact on their children, but they recognized that it was difficult to make such abstract concerns about the future part of anyone's daily lived experience. So they launched DearTomorrow, which invites people from around the world to write private or public letters to their children, grandchildren, or future selves—to be opened in the year 2050. Their aim is to spread the practice, and the effort has won several awards.

The audience of someone living fifty years from now, especially if it is a child, is an anchor in the future to which people can tie the listless boats of their attention in the present. Writing a letter to the future projects the imagination ahead in a concrete way, to clarify the potential consequences of choices in the present. Shrum and Kubit plan to study whether such letter writing changes people's current choices and behavior. I adapted their technique recently for my own purposes, writing a letter to myself thirty years in the future, in which I imagined the outcome of having made divergent decisions in the present. It was an even more powerful experience than seeing an aged photograph of myself.

Within families, we can adopt rituals to envision the future together. A few years ago, Seattle-based entrepreneur Michael Hebb did this when he showed up in a coffin to his fortieth-birthday dinner. Hebb, who lost his father when he was a teenager to Alzheimer's, wanted to invite people in his life, including his children, to talk about his eventual death, and for people to be able to share what they hoped for their own deaths while they were still alive. While seeing a gravestone is solitary, Hebb created a collective practice to get people to talk about the inevitable future of aging and dying with their families. He published a free playbook for holding such dinner parties, with elements whimsical and serious, and now estimates that more than a hundred thousand people have held the dinners in dozens of countries. He calls the movement "Death Over Dinner." An acquaintance practices a similar ritual with friends—an annual gathering where each person writes his own obituary and reads it aloud to the group. It is a yearly rite of projecting oneself across the gravestone's dash, between birth and death.

t might seem fairly simple to some people to imagine their own future, or even the later lives of their children. For the truly ambitious, it is also possible to summon the distant future—what lies well beyond everyday contemplation—in the present.

In an unwelcoming expanse of the Chihuahuan Desert in southern Texas lies a ghost meant to stretch the imagination like this, far past a lifetime. Danny Hillis, an engineer who forged some of the world's most advanced technology, is conjuring the ghost.

While he was a doctoral student at MIT in the early 1980s, Hillis invented one of the world's fastest computers. Called the Connection Machine, the parallel supercomputer could make multiple calculations simultaneously, and he fashioned it as a tool to build artificial intelligence. (His company's motto was "We're building a machine that will be proud of us.") Similar technology is still being used today in supercomputers that simulate neural circuits in the human brain and predict complex weather patterns. In the '90s, Hillis went to work for Disney, heading its Imagineering studio alongside Bran Ferren and inventing multimedia theme park rides and museum exhibitions around the world.

Hillis grew up on Space Age science fiction, rife with utopic visions of future explorations of the universe and technological feats. But in the 1990s, he began to feel that people were not thinking or dreaming about the future anymore. "The millennium became a barrier for people— they stopped thinking past the year 2000," he reflected on a summer day as we sat in his backyard. He felt his own future was shrinking by one year with every passing year of his life. He longed to feel some greater meaning across time, and to activate his own imagination. So he began to dream of creating something that would transport his mind to the far-off future—a feat of engineering that had never been attempted before. His idea was to build a landscape-scale clock that would run for ten thousand years.

The clock began as a fantasy but is now being built in a five-hundred-foot-tall limestone cliff on a tract of Texas land owned by the clock's benefactor, Jeff Bezos, the founder of Amazon. There, sharp-tongued cacti and brambly greasewood shrubs crowd in valley thickets, and low mountains rise from saw-toothed rock, into which no paths have been hewn. The desert climate in Hillis's mind creates the possibility for the clock to survive periods of neglect. That it lies hidden from view, in an expanse of uninviting private property, cannot hurt.

To create the clock chamber, Hillis and an unlikely team of miners and wine cave designers blasted a tunnel into the cliff with explosives. A robotic diamond drill carved a staircase of 365 steps, one for each day of the year, that will descend to the clock.

The machine will have to run far more slowly and for far longer than any of Hillis's supercomputers or other inventions. The clock bearings will be made of ceramic, its windows forged in sapphire. It will measure time not in the milliseconds that define the speed of supercomputers but in the increments of years, centuries, and millennia.

The clock's pendulum will be powered by the temperature differences between day and night. (It will be calibrated each year on the summer solstice, by the sunlight entering the chamber through an opening.) Hillis envisions that visitors to the clock, whenever they arrive, will wind it up with a wheel to show the date and time of day on the face, as an arrangement of celestial bodies—the position of stars in the night sky, the sun, and phases of the moon as viewed from Earth.

Building this clock, Hillis has found, gives him an excuse to think about the distant future. To contemplate the details of how to make a machine run for that long is an exercise in imagining what could change over the life span of the clock. Over the course of ten thousand years, for example, Earth is likely to sustain a volcanic eruption powerful enough to shroud the sun. The pendulum of the clock has to be able to store enough energy to keep ticking even if there is a century of darkness. Languages are likely to change, and so he and his collaborators have developed a "Rosetta disk" to leave in multiple copies around the planet, as a decoder ring for future visitors. For Hillis, thinking about the clock has made abstract problems like climate change feel more real. He is building something concrete that has to endure the changes on the planet that he will not witness himself, and that his kids will not witness, either.

The clock idea captivated those whom Hillis told about it, including futurist and technology guru Stewart Brand and the musician Brian Eno. The group put together a foundation called Long Now to oversee the clock's stewardship over time, which has recruited more than eight thousand enthusiasts to its ranks. Long Now members, engaged in collective imagination of the next ten thousand years, are a kind of subculture or, as one member described it to me, "a support group for people who think about the future." Some ask, How do we create a legal framework that can last for ten thousand years? Others ponder the weather or focus on the engineering problems. The clock lets them picture the unknowable.

Around the world, other groups are setting anchors for the imagination in the distant future. In Halberstadt, Germany, the St. Burchardi Church has rigged its organ to play a John Cage piece with ambiguous time signature, *As Slow as Possible*, at a pace that will last more than six hundred years. The organ will be built and disassembled as the piece goes on, to add and subtract pipes for the notes to be struck next. Notes sound twice a year—at most. In a sense, the musical piece is a collaboration across generations, and an invitation to maintain the relevance of an institution and an instrument over time.

These quirky experiments call to mind the Voyager capsules launched into the outer solar system in 1977 to communicate with potential aliens. The capsules carried golden records with audio clips of laughing children, elephant calls, Navajo chants, Bach concertos, and a fighter jet. The records bore images of human sex organs, a mountain climber, and the Taj Mahal. People around the world were enthralled by the artifacts sent from our planet in an attempt to contact life on distant planets. The astronomer Carl Sagan, who led the effort, said the project's grander aim was to help define humanity and human civilization for those on Earth. Today, the clock and organ experiments—whether or not they persist in the long run—are perhaps most powerful in helping people here and now call forth the far-off future, and see ourselves as part of a vast span of time.

Even without the budget or the drive to build a massive clock in the desert, each of us can find ways to imagine the future better—whether by writing letters to our future selves or hypothetical great-grandchildren, drafting our own obituaries, or creating something that we want to last beyond our lifetimes. The steps might be as simple as planting a tree in a neighborhood, leaving books to a library, or starting a perennial garden—but they can be powerful practices to inhabit our imaginations of the future. We can also leave our minds free to wander to the future more often, so that its possibilities, good and bad, become more real to us. Most people do not have time to do this every waking hour of every day, but can still dedicate some share of time—a day a month, an hour a week—to our future selves. We might even lift our gaze to the road ahead by conjuring future episodes as if they are memories, seeing ourselves acting in them and getting through the thickets.

The challenge we face is that even with the power to imagine the future, it is difficult to overcome the demands of the present.

2

DASHBOARD DRIVING

Counting What Matters

"Dry again?" said the Crab to the Rock-Pool.
"So would you be," replied the Rock-Pool,
"if you had to satisfy, twice a day, the
insatiable sea."

—CYRIL CONNOLLY, *The Unquiet Grave*

A friend once caught the fever of counting her steps. She was taking part in a workplace fitness competition, in the hope that it would cajole her into getting healthier. Her company doled out Fitbits— those digital schoolmarms that ceaselessly track their wearer's every move—to teams of employees who competed against each other.

One day I ran into her in Kendall Square, the bustling technology hub of Cambridge, Massachusetts, and asked her how she was faring. She had found her team lagging behind the others, so she had added a short afternoon walk to her daily routine. But she confessed that her walks to ratchet up her number of steps took her past a delicious bakery, where she invariably bought herself a pastry or two that she gobbled down on her way back to the office. She had gained a few pounds since the competition began, but her team had broken into the lead. A measure of progress, the number of steps she took each day, had replaced the goal of actually being fit. It contributed to an illusion of health, even though she had overloaded her diet with sugar.

Today we are awash in sensors and devices that allow us to track nearly every aspect of our lives and measure our progress continuously, in real time. Technologists are calling data, in turns, the oil and the electricity of the twenty-first century. Athletes chart their resting heart rates. Commodity traders detect tiny fluctuations in weather forecasts. The unhappy track their momentary mood changes. Parents monitor their babies' sleep habits and diaper dampness.

By 2020, we'll have tens of billions of sensors on Earth, allowing us to mince and measure yet more of our everyday actions. The omnipresence of digital measuring sticks makes it possible to take stock with greater and greater granularity and frequency. Yet technology is only accelerating the pace of a long cultural trend of measuring in smaller increments. The writer Dan Falk points out that in Shakespeare's time, a "moment" could mean an hour, and that the Bard never mentioned the unit of a second in any of his writing. Chaucer, he notes, lacked the concept of the length of a minute.

The upside of so much measuring is that it lets us see where we stand. Like vigilant watchmen, data points we collect tell us whether we are slipping off the path of progress. High temperature readings, for instance, tell us when to take Tylenol or call the doctor. Measuring also gives us the opportunity to meet incremental targets toward daunting goals. Counting every step can motivate people to exercise more, encouraging us to use ten thousand paces as a daily dartboard. We take stock of earnings, page views, and test scores so we can strive to improve them, in an attempt to make a little more money, have a few more readers, get into the college we want—in short, to win at the game of life. As we gather more data, we come to rely more on metrics—the close-range targets that reflect what we can best measure—as bellwethers of later success or harbingers of future danger. It all seems helpful, or at least harmless.

Vijay Mahajan began his eighty-day pilgrimage across India at the beginning of 2011 with a burning question: What had gone wrong?

More to the point, he wondered whether the past thirty years of his life had been a waste. He had devoted much of his career to serving the poor, building an industry intent on breaking the cycle of poverty. Now he stood in the Sevagram Ashram, a collection of modest cottages in central India where Mahatma Gandhi had planned India's peaceful revolution in the late 1930s and early 1940s. Sprawling mango and fig trees shaded the courtyard and prayer ground. It was a place Mahajan had visited three decades before, as a young and idealistic man, well before the events that shattered him.

He staggered, already tired on the first day of his journey, to the bamboo-and-palm-roofed cottage where Gandhi had met with the visitors who came to see him from all over the world. Engraved on the entrance gate were what Gandhi deemed the seven social sins, among them "Commerce without Morality."

Long before his pilgrimage, Mahajan had founded one of the first organizations in India to offer small loans to the country's rural poor and urban slum dwellers. The service, known as microcredit, helped families and farmers buy livestock and start small businesses with relatively small sums—sometimes as little as a hundred dollars.

In the decades that followed, Indians became the world's leading recipients of microcredit. Companies like Mahajan's extended loans to millions of people who had lacked access to traditional banking and formal lenders in the past. He was widely hailed as a pioneer of the booming Indian microfinance industry, championing the cause publicly and globally. In 2002, the Schwab Foundation selected Mahajan as one of sixty outstanding social entrepreneurs in the world. In 2009, *BusinessWeek* named him one of India's fifty most powerful people.

By the fall of 2010, however, the industry once lauded as a savior, and its foremost champion, were being painted as villains. A rash of suicides erupted among villagers in Andhra Pradesh, a southern state with a population the size of the entire country of Colombia. Villagers told stories of being bullied by microloan officers. Hundreds of people killed themselves, citing their shame and desperation at not being able to repay their loans from microlenders. The state government blamed a reckless industry and used the incidents as justification to banish private lenders, including Mahajan's company. An ordinance instructed residents not to repay their loans.

I followed closely the news of this crisis as it unfolded in the country where my parents were born. It seemed needlessly tragic, and it unsettled me. I had had the impression that microfinance was an ingenious way to help the poor, filling their modest but neglected need for credit. A few years later, I tried to better understand what had happened. That's when I met Vijay Mahajan.

At the time of the crisis, he told me, India's leading microfinance companies had concentrated the majority of their loan portfolios in Andhra Pradesh. The state government response nearly destroyed BASIX, Mahajan's company, and shook the entire Indian microfinance industry. BASIX lost more than a million credit customers, and Mahajan laid off nine thousand people, mostly loan officers who worked in the field. He felt that his model of rural development had been different from that of other private microfinance lenders, but that seemed to matter little in the end. He'd been nearly wiped out along with the rest of the lenders in the state, and country. The company was more than $450 million in debt.

At that point, Mahajan began to wonder whether his work had done more harm to the poor than good. How had he and the industry he had built fallen from such great heights?

In January 2011, on the sixty-third anniversary of the assassination of Mahatma Gandhi, Mahajan set out on a Shodh Yatra, a journey in search of truth. "What is real is what the people say," he wrote in his journal. Over the following months, he covered more than three thousand miles of terrain, visiting ancient shrines, roadside recycling sheds, flower vendors' stalls, a blind man's phone booth. He walked and drove along

India's back roads, within its villages and slums and farms, to learn what the poor people of his country had to teach him.

Vijay Mahajan began his journey when he was fifty-seven years old. Over the course of his trip, a salt-and-pepper beard emerged on the lower half of his face while his hairline receded. On most days of the pilgrimage, he wore a simple cotton kurta that tightened over the paunch of his midsection and fell below his knees.

Mahajan has a careful, eloquent way of speaking, and a presence that commands attention. He strikes a strange chord as a businessman: He is as comfortable parsing Yeats and Auden as he is parsing financial models. I was fascinated when we first spoke in 2016 by his penchant for self-reflection in a situation where many people would choose denial or self-justification. Before the crisis struck Andhra Pradesh and the microfinance industry, Mahajan had been contemplating an early retirement. In the aftermath, he felt he could not just walk away.

Mahajan was born in 1954 in Pune, a western city flanked by emerald hills, dubbed the "Oxford of the East" because of its universities. He was the fourth son of an officer in the Indian army. All of Vijay's three older brothers followed the patriarch into military service. As his parents became rooted in the middle class, Vijay was sent to an English-language school, where he learned among the Jesuits at St. Xavier's in Jaipur, home to India's magnificent palaces, the legacies of maharajas. He cultivated a love of English poetry and literature alongside technical interests, and was a star student, selected for admission into the elite Indian Institute of Technology.

He entered adulthood during the tumultuous 1970s, when student activists protested rising food prices in Gujarat (in northwestern India) and a sweeping movement rose up in Bihar (in northeastern India) in response to government corruption. Prime Minister Indira Gandhi's government cracked down on these social movements as they spread, curbing civil liberties and arresting opposition leaders. Mahajan saw young people emerge as leaders amid this turmoil, fighting on behalf of the poor and powerless. He became conscious of a desire to dedicate his own life to the cause.

For years, however, he hesitated. He was drawn to more stable and lucrative careers in private industry and compelled by ambition to earn a graduate degree in business management. Looking back, he compares himself at that age to Jean-Paul Sartre's protagonist Mathieu in the Roads to Freedom trilogy, who at first indulged the convenience of believing that a laudable ideology was enough, instead of plunging himself into the anti-fascist resistance movement of World War II Europe. Mahajan soon realized that he had to act on his ideals to build his commitment to them. Meeting his wife-to-be, Savita, in graduate school, who encouraged him to work in public service, sealed his fate.

In the early 1980s, fresh out of graduate school, Vijay began a career working on behalf of the poor. A nonprofit group, the Association for Sarva Seva Farms, recruited him to the rural outskirts of the state of Bihar in northeast India. The work built on the legacy of Vinoba Bhave, an acolyte of Mahatma Gandhi who had walked across India in the '50s and '60s to persuade wealthy landowners to donate some of their land to the poor. As a result of his Bhoodan movement, more than 2.4 million acres were given to landless poor people to cultivate. In many areas of India, including Bihar, however, the land remained marginal even into the '80s. Mahajan worked literally hand in hand with rural people to level their land, remove boulders, and dig borewells for irrigation.

He soon encountered problems among the poor that ran far deeper than whether their land was arable. They often had entrepreneurial ambitions but no training in how to start or run a business. They lacked formal ways to save money in bank accounts or to get credit. A few years later, Mahajan founded a nonprofit that recruited professionals—including businessmen, doctors, and veterinarians—to help rural people start their own enterprises. He tried to persuade local banks to offer loans for the purpose of starting farms and small businesses but found that traditional bankers were unwilling to lend to people who had no collateral and no credit history.

In 1994, Mahajan learned about the work of Muhammad Yunus in Bangladesh. Yunus, an economics professor, had witnessed the effects of the famine that struck his country in the 1970s, and began lending small amounts of his own money to poor women living in villages. He found that with small loans, the women could grow their businesses making bamboo furniture—and that they reliably repaid his loans, on time. Seeing a gap, Yunus founded his own financial institution, the Grameen Bank, which has been wildly influential, spawning similar banks and companies all over the world. In 2006, Yunus won the Nobel Peace Prize.

Inspired by Yunus, Mahajan decided to launch a similar organization in India that would lend to the poor. In 1996, he founded BASIX. Its first loans were hard won. Mahajan had to persuade the Reserve Bank of India to let banks lend to his company so that it could in turn lend to poor customers. He also had to raise philanthropic donations and foreign aid, because he and his cofounders did not have enough capital to support the lending on their own.

The efforts paid off. Other lenders began to enter the Indian market offering microloans. As people repaid their loans, the capital base grew and it became possible to extend credit to more and more people.

By the early 2000s, however, Mahajan had grown weary of panhandling for donated funds to operate. He saw the potential of wedding a for-profit enterprise of microlending with his broader social mission. So he turned BASIX into a holding company with a for-profit arm that offered microcredit, allowing it to take on foreign equity investors. That arm subsidized other work that was less lucrative, which BASIX also provided to the rural poor, such as rainfall and crop insurance, vaccinations for farm animals, business training, and savings programs.

Anyone who has visited India has had their senses assaulted by its bold colors, intense smells, and persistent soundtrack of deafening and melodic sounds. Marigold garlands hang from honking auto rickshaws. Traffic stops for turmeric-stained bulls. The spiced aromas of roadside kitchens waft into the nose along with diesel fumes, while bhangra beats thunder from car stereos.

Nothing hits me as hard on my travels there, however, as the overwhelming presence of poverty. The cupped hands of kids who looked like me will be forever seared into my childhood memories. For millennia, poverty in India has been passed like a curse within families from generation to generation. For much of the country's history, its most destitute people lacked access to banking—the ability to save, invest, and borrow money from formal lenders—which reinforced their poverty. When a poor person is desperate to pay a medical bill or to buy food for her family and has no reliable source of cash, pawnbrokers and local godfathers, with no curbs on their collection practices or interest rates, often fill the vacuum.

Over the years, the Indian government has made modest attempts to fix the problem. Efforts to create cooperative credit for the poor date back to when India was part of the British Empire. After the country won independence, India's banks, under the control of the national government, extended such programs to reach more rural people.

In the 1980s, aided by nonprofit organizations and the government, rural women in India began to form cooperatives, known as self-help groups, to gain access to such credit—money that could help them cover household expenses. These groups were typically composed of ten to twenty women in a village who had regular meetings and would help one another repay loans—reducing the risk for the lenders. With policies that made the poor a priority, the Indian government began to back these women's groups, linking them to national banks and international development banks. Still, much of the rural poor did not manage to secure formal loans under this system.

When microfinance groups like BASIX, inspired by the Grameen Bank, first sprang up in India, they were nonprofit organizations. The self-help groups were natural targets and allies. Women in the developing world, researchers have shown, often want to start businesses and earn their own living. And they typically put money they earn back into their households and businesses, helping their families escape the cycle of poverty. The self-help groups also understood what taking on a loan required of them. The women covered one another's payments in a pinch, and they held one another accountable.

The microfinance industry in India, and in particular in Andhra Pradesh, grew rapidly, and was booming on the eve of the 2010 crisis. As for-profit companies, the major Indian microlenders had recruited private venture capital investors from abroad, including the United States. More than $500 million flowed to the sector between April 2008 and July 2010. The lenders attracted these investors with the promise—and delivery—of astronomically high returns on their investments. High repayment rates on loans—with less than 3 percent of poor borrowers defaulting—signaled stability. A growing book of borrowers signaled growth.

In 2008 and 2009, the ten largest microfinance companies in India earned, on average, returns on equity of more than 35 percent for their investors. Some companies aimed to go public, to become traded on the stock market and access even vaster stores of capital. The higher their value when that happened, the more their early investors would cash in. To boost their estimated value for their initial public offerings, several companies sought to grow the number of loans in their portfolio without increasing costs. The easiest way to do that was for companies to increase the amount of lending in regions where they already had lots of borrowers and hence field officers to sign people up for loans—in other words, to sign more poor people up for loans in places like Andhra Pradesh.

Amid this fever, many microfinance loan officers, young men from rural communities themselves, had marching orders from their companies to sign as many loans as possible in the state—some with the incentive of 100 percent salary increases or free motorcycles and televisions for getting more loans on the books in a given week. The companies paid special attention to the women's self-help groups. The officers had the motivation to hit near-term targets—just like my friend with her Fitbit.

In 2009, the state was home to only 7 percent of the country's population and yet accounted for 30 percent of its microloan portfolio. By

2010, there were about six million borrowers in Andhra Pradesh who had taken on nine million loans—which meant many poor people had multiple sources of debt.

The fastest and most aggressively growing Indian microfinance company at the time was SKS Microfinance Limited. Its founder, Vikram Akula, was something of a wunderkind and media darling. By age thirty-seven, he had already been on *Time*'s list of the hundred most influential people and profiled by major news organizations in the United States and India, and was considered, like Mahajan, a leading figure in the global microfinance industry.

On the eve of the crisis, Akula was taking his company public. He had raised more than $150 million in equity capital from venture firms, including Sequoia Capital and Sandstone Capital in the United States—investors that had seen extraordinary returns and now expected to make significant profits.

Between April 2008 and March 2010, SKS added more than four million borrowers to its books, which amounted to 488 loans per officer. It's impossible to imagine that a loan officer could have personally known his clients, and whether they were qualified to get loans or could afford to repay them. In July 2010, just months before the crisis, SKS went public. It was valued at $1.5 billion—forty times its earnings that year.

Akula is a controversial figure in India and in the microfinance industry. He once said of the business of lending to the country's poorest people, "This work can be driven only by greed." In confidence, his critics compare him to Gordon Gekko, the character Michael Douglas played in Oliver Stone's 1987 film *Wall Street*, who proclaimed that "greed, for lack of a better word, is good."

I tried to interview Akula for this book, but after a few e-mail exchanges he seemed to shy away from having a conversation. A friend told me he was reticent—understandably, perhaps—to talk to the press after how he was portrayed after the crisis.

Muhammad Yunus, the founder of the Grameen Bank in Bangladesh and the modern microfinance movement, openly criticized Akula's model of making huge profits while serving the poor. As microfinance lenders around the world became for-profit companies in the nineties and early aughts, Yunus worried that they would stray from their core mission. In a 2009 interview with *Forbes India*, he warned of "loan sharks under the guise" of microfinance lenders. He openly debated Akula in a 2010 meeting before the crisis. But the for-profit model attracted deep-pocketed and influential Silicon Valley investors who saw microfinance as supporting entrepreneurs—something that aligned with their own credo of how to save the world, even if it did not always align with the reality of why poor people took on loans.

All eyes in the microfinance field were fixed on India's manic growth during that period. Elisabeth Rhyne, an expert who led microfinance programs at the U.S. Agency for International Development and Accion International for more than a quarter century, recalls that at every international development conference in the late aughts, people in the microfinance field could not help but talk about the astounding numbers in India—at how many poor people were being helped. Microlending companies overlooked potential risks, Rhyne says, and so did the Indian banking authorities setting policy for lending and much of the global microfinance community.

A quiet debate was emerging, however, among some in the global microfinance industry as Indian companies looked to go public. Was the skyrocketing growth in the number of loans in India due to the vast unmet needs of poor people? Or was something else happening?

Hitting a near-term target can become an obsession. We often set these targets, known as metrics, based on what we can easily measure. We believe that numbers are objective—that numbers don't lie. And so we use them to determine whether we are failing or succeeding.

We tend to rely on short-term data points because they align with what we see in front of us. In other words, they reinforce the availability bias—our tendency to take most seriously what we can take in or conjure now with our senses. At times, however, close-range measures outright deceive us. I shivered through record-breaking cold days while living in Washington, D.C., in the winter of 2014, when icy Arctic winds were unleashed from what meteorologists dubbed the "polar vortex." But the thermometer readings on those mornings belied the larger trend—on balance, it was the warmest year ever recorded on the planet. Temporary dips in gas prices the following year sent Americans on buying sprees for SUVs, condemning them to higher expenses when prices would invariably rise again.

For similar reasons, getting a $2,000 check in the mail can make people feel like a tax policy, such as that passed in the United States in December 2017, is a good idea, even if that same policy makes it more difficult for the middle class to buy homes, pay off debts, and go to college—denying them far greater wealth over time. Politicians get away with it; there's little public outrage. The check we get this year—and the uptick of our bank balances—is a blindfold, more salient than the overall eventual loss we'll experience from the tax reforms. It's much harder to see the larger trends not captured in a single data point or a sum we get right away. Taking stock now, in this sense, obscures our view of future threats.

The numeric targets we choose shape not just our perspective but also our actions. The old adage "What gets measured gets done" ought to be flipped today: We do whatever we can measure. I call this problem dashboard driving, because we careen forward glued to gauges of speed or fuel levels, not realizing meanwhile that we may be steering off a cliff. It's not simply our nature that drives us to favor the present and neglect the future; it's the tools we choose to use, our ostensible measures of progress. The upside of this insight is that we do have a choice.

Studies of New York City cabdrivers illustrate how metrics shape everyday decisions. A cabdriver will typically set monthly or yearly earnings targets. But several research projects have shown that on a given day, most drivers will make choices based on daily metrics, even when they are at odds with their own long-term goals. One group of economists found, for example, that cabdrivers who achieve their informal daily target for earnings on a rainy day will quit hours early, even though on such days there is greater potential for them to earn more by working longer hours because more people want to take taxis. On days with nicer weather, when it would make sense to take time off, drivers instead work longer hours, cruising around to find customers, wasting their time and fuel. Their daily target, albeit implicit, seems to guide them more than their explicit aims for the future. It creates a sense of accomplishment that reinforces the choice to neglect the future consequences of quitting for the day. And it is measurable in the moment.

It might seem foolish to be so easily swayed by the immediate and ephemeral. But one reason people take the bait of a close-range target is that we try to avoid immediate loss. Most of us hate to lose even more than we crave winning.

To lose something, however trivial, summons the uncomfortable feeling of lacking control. Money forgone or a target missed today stands for something more than that specific loss; it confronts us with a reality that contradicts our expectations. Losing inflicts the emotional pain of contemplating that we may not be as good as we think or may have made a mistake. Perhaps we try to avoid missing targets because it's like losing a grip on the rope we are using to climb a steep mountain, reminding us how vulnerable we are to gravity.

Daniel Kahneman argues that our aversion to losing has roots in the early stages of human evolution. To avoid predators, hunter-gatherers urgently responded to threats, he writes in *Thinking Fast and Slow*, and the people who did that successfully survived and reproduced, passing along the trait. As a result, the modern human has inherited the impulse to protect herself from loss more aggressively than she seeks gains. In experiments he conducted with his longtime collaborator Amos Tversky, Kahneman showed how the emotional prospect of an immediate loss affects people's decisions far more than a view of what they are likely to win or lose in the long run. This means that people are predisposed to weighing a short-term sacrifice—or sign of loss—too heavily when comparing it with a sign of future opportunity or danger. The taxi drivers who are just short of missing a daily target may act to avoid feeling such loss.

In 2009, more than a year before the crackdown in Andhra Pradesh, a front-page story in *The Wall Street Journal* drew on voices from the microfinance industry in India, Europe, and the United States who warned that the region had a credit crisis in the making.

An expert from the Indian Institute of Management said that lenders were "carpet bombing" particular neighborhoods and regions with loans. The manager of a $100 million investment fund expressed fear that a bubble was swelling. In a series of articles, journalist Ketaki Gokhale reported on Indian women who had taken on several microloans and were struggling to repay them. One had taken out loans with nine different lenders and felt she was being publicly shamed for falling behind on payments. Others had been offered loans even though their incomes were as little as nine dollars a month.

Gokhale found women who were not starting businesses with their loans, but rather using them to buy milk, pay overdue bills, and cover relatives' wedding expenses. The loans, in these instances, were not helping women raise their incomes, but rather sticking temporary Band-Aids on bad financial situations. With their high interest rates, the loans might have even been digging these women deeper into poverty.

Daniel Rozas, a consultant to the microfinance industry generally supportive of and sympathetic to its aims, warned of a credit bubble in the Indian microfinance industry unless businesses changed their practices.

Rozas was not clairvoyant. Nor was he relying on any insider knowledge or elusive data. He was simply looking beyond the metric of the high repayment rates on loans, a measure being summoned by microfinance lenders as a sign that all was well in Andhra Pradesh.

Rozas is an American expat based in Brussels, and we spoke by phone in 2016. He told me that the capacity of the state—the number of potential borrowers, depending on its population and other factors—in 2009 seemed outstripped. He made some coarse, back-of-the-napkin-style calculations at the time and concluded that it was not at all reassuring that people were repaying loans. It could mean that they were doing so by taking on yet more loans pushed on them by eager microfinance field officers, not by generating income through new businesses. At some point, when people could not get any more credit to pay back their high-interest loans to the lenders, the house of cards would collapse. Families would be drowning in debt they could not repay. Companies would fail.

Rozas had spent the previous chapter of his career at Fannie Mae, the mortgage financier, where he had a front-row seat as the U.S. mortgage market collapsed in 2007, spawning the chain reaction that led to the most devastating global financial crisis since the Great Depression. The housing market bubble had grown with the influx of Wall Street capital, which supported rabid underwriting by mortgage lenders. Some of them had predatory practices and products requiring borrowers to refinance regularly and to face rising unpaid balances over time. Firms like Lehman Brothers and Bear Stearns packaged the high-risk loans into securities and then sold off the risky bundles to often unwitting investors. Credit agencies failed to rate the investments as risky. When the housing market became oversaturated and homeowners could neither sell their homes nor make their payments, dominoes fell throughout the financial system. Fannie Mae, despite having relatively few of the pure subprime mortgages in its portfolio, had nonetheless taken on many risky loans in an attempt not to get squeezed out of the booming market. When the bubble burst, the company got crushed, and, along with Freddie Mac, was bailed out by U.S. taxpayers.

"Frankly, the numbers there concern me," Rozas wrote in a commentary he published in 2009 about Andhra Pradesh. He believed with near certainty that the manic growth of microfinance loans in the state was the result of borrowers' taking on too many loans. "I think this is about the strongest evidence of a bubble one could hope to find using publicly available data."

To continue to build up microfinance debt among India's southern poor who already had loans, Rozas warned, "puts short-term gain not only above the long-term financial soundness of the sector, but, more importantly, above the long-term interests of the very poor [they] are seeking to serve."

Vikram Akula, on the other hand, responded to warnings of a credit crisis with a vigorous hand wave.

In 2009, in a *Harvard Business Review* commentary and a published letter to the editor of *The Wall Street Journal*, Akula dismissed those who argued that a microcredit bubble was about to burst in India. He pointed to what he saw as the key signs of the industry's health and stability. First among them was the high loan repayment rates—at his company 99 percent, and for the Indian microfinance industry as a whole nearly 98 percent. Surely these measures showed that the rapid growth would continue.

In response to criticism that too many borrowers were taking on multiple loans, Akula leaned again on repayment rates. He pointed to a research paper that had shown that certain ambitious and motivated entrepreneurs borrowed from multiple microlending companies to grow their businesses, and repaid their loans at a high rate. What could be bad about that?

Rozas, however, looking more deeply at that paper and the metric of the repayment rate, found that it did not apply to the situation that was unfolding in 2009, but rather to an earlier phase before the rapid growth of microfinance in India. It was no longer just a small group of enterprising standouts taking on multiple loans, but instead broad swaths of borrowers—many of whom were likely not using the loans to start businesses, but to deal with family medical emergencies and put food on their tables.

The crash came on like a rogue wave. Under the surface, however, the pressure had been building for months. Borrowers took on multiple loans that they could not repay. Ashamed villagers across the state of

Andhra Pradesh, harangued by neighbors or loan officers, killed themselves. The state government seized upon the suicide reports, kicked out the private lenders, and made it legal for borrowers to leave their debts unpaid. That triggered the collapse, wiping out entire companies and plunging those that remained into deep debt.

People sympathetic to Akula say he became misguided by the outsize success he was experiencing—and the pressures of having taken money from investors whose focus was exiting with high returns when the company went public, not the long-term viability of the business or the future of serving the poor with loans.

I believe it would have been difficult for Akula to say at the height of the bubble—even if he suspected it to be true—that the industry was growing too fast. It's a rare corporate board that wants to hear that it has to slow down growth, or that its shares' value should not be maximized when going public. Rather, it's the kind of statement that gets CEOs fired. Or, as Chuck Prince, former CEO of Citigroup, Inc., notoriously told a reporter from the *Financial Times*, "As long as the music is playing, you've got to get up and dance."

I talked to several people who felt Akula's attitude in the lead-up to the crisis was too brazen—including experts who did not face his constraints of meeting investors' outsize expectations. When I asked why they had not been more vocal about their skepticism, each said they had trouble openly criticizing him because of how good the numbers looked.

The measure hid the impending danger.

The ancient Greek historian Herodotus wrote about the wise statesman Solon, whose election as chief magistrate of Athens in 594 BCE led to reforms including a ban on the practice of enslaving people for unpaid debt, and suffrage for plebeians. After passing his reforms, Solon voluntarily fled Athens. He traveled to Sardis, in modern-day Turkey, to visit King Croesus.

The vain king promptly sent his guest Solon on a tour of his palace to size up his riches, according to Herodotus. The king then asked the wise man to name the happiest man. You might call it leading the witness.

Solon angered Croesus by not naming the king himself. The happiest man, Solon said, was an Athenian named Tellus, who died heroically in battle. All of his grandchildren survived him and he was honored after his death for his feats. He advised the king that the measure of a fortune at a given time in a man's life is not a true measure of whether his life is fulfilling in total. An average lifetime lasts 26,250 days, Solon said, and misfortune could befall a person on any one of them. (And misfortune did befall King Croesus, who lost both his son and his kingdom.)

More than two centuries later, Aristotle echoed the wisdom of Solon in his treatise on ethics, arguing that a life ought not be measured in small increments but rather in its sum, viewed from a long perspective. When we measure ourselves only by what we achieve in the immediate, we inhibit our patience for endeavors that take a long time to bear fruit, whether learning a new language or raising a child. We also fail to see the big picture, the larger meaning of our decisions as they play out over time.

The insight seems to be both timeless and perennially overlooked. When Thomas Mann wrote his epic novel *The Magic Mountain* in the 1920s, he seemed to be reflecting on the vanity of frequent measurement. In the story, patients cut off from society at the Berghof sanatorium in alpine Switzerland are instructed by medical staff to take their temperature four times a day in seven-minute increments. The readings give a tempo to each day and a report card to the patients on their performance. Meanwhile, the same patients appear oblivious to the passing of months and years leading up to World War I. The act of measuring seems to reinforce their feelings of illness, and thermometers are carried as badges of honor by the infirm that the healthy are not deserving of. When the protagonist, Hans Castorp, proposes to leave the sanatorium, he asks the doctor whether he is well enough to do so, given his temperature chart. The doctor reveals in anger that the thermometer readings were meaningless from the start, a fiction that signaled nothing about the patient's condition. Meanwhile, the temperature taking served as false reassurance that hiding out from a society fraying into conflict was time well spent.

No one thinks: "I'm going to choose a myopic metric." In fact, we often choose to measure ourselves against close-range targets because of a belief that it will actually *help* us with valuing future consequences in our current decisions. We measure what we do now because we want to know where we stand—and the anxiety of not knowing can be great. People, organizations, and societies often adopt a metric because they believe it is a proxy for a deeper, delayed outcome. It's the way we glue our gaze to the metrics that causes the problems.

One way we can resist becoming myopic is to tune out the noise of near-term metrics when they are distracting us from the big picture.

A hedge fund investor from Chicago, Anne Dias, once told me about how she does this. She knew that investors who look regularly at their portfolio earnings and losses throughout the day amass lower profits than those who look less often at how they are faring. Aversion to loss drives investors to panic and make rash decisions in the moment to sell off investments on the basis of what might just be temporary dips in value— even when those very stocks will be more valuable over time. University of California economists Brad Barber and Terrance Odean studied this phenomenon and found that thousands of investors hold on to stocks that underperform the market and sell stocks that outperform the market because of their emotional impulses to avoid immediate loss. (Billionaire investor Warren Buffett has similarly said that he often makes more money when he is sleeping than when he is actively investing.)

Dias decided to shield herself from looking at her investment portfolio too often and asked her staff to tell her only when it dipped or rose past a certain threshold. In effect, she created a bubble to protect herself from acting on her tendency to dodge loss. She bound her future self to patience with a decision she made in the present.

Tuning out noise is a practice we can each adopt in our lives, especially when making important decisions. When buying a house, for example, we might focus less on calculating near-term payments and more on the potential costs over the decades we expect to live in the home. When buying a car with low fuel efficiency, we might consider expenses not just based on this year's gas prices, but looking at many years on average. When we have long-term projects we want to tackle, we might avoid frequently looking at our e-mail in-boxes and social media feeds—spacing out the increments by which we view how many messages have gone unanswered so that we measure ourselves by the amount of work we have done toward long-range goals, instead of how far we are ahead or behind on today's communication.

Like Dias, we can bind ourselves in prescient moments—those when we can exercise foresight—to future actions we want to either take or avoid. This can keep us from overreacting when we miss—or appear near to missing—a target, shielding us from the magnetism of one metric. The techniques for doing this can take the form of contracts we make with ourselves, with consequences for reneging. We can also consider replacing single metrics with many measures of progress when we feel the need to know—or be motivated by—how much we've done each day or week. Instead of just our number of steps each day, for example, we might also measure the pounds we lose each month, the calories we eat or burn each week, how we are feeling in terms of strength and fitness, and how fast we can walk for how long. We can also wait to act on what we learn from a measurement—and remind ourselves of the option of not reacting at all.

During India's microfinance boom, the prevailing metric of the loan repayment rate told a deceptive story. It did not measure the health of the industry, or its future prospects in Andhra Pradesh. It did not measure improvement in the lives of the poor. Meanwhile, the high repayment rate created an illusion for Vijay Mahajan. It obscured threats to his company, his industry, the poor families who took on loans, and his reputation.

Mahajan was not completely oblivious to the problems afoot, however. Before the crisis, he had a sense that the industry needed to rein in its lending practices. Less than a year before the Andhra crackdown, along with his industry colleagues, he started a network of forty-four microfinance lenders. Mahajan took the helm as president of the Microfinance Institutions Network and set norms for the lenders, including the precept of not issuing more than three microfinance loans to a single borrower.

Not everyone in the industry put these voluntary norms into practice, however, and rules were not enforced. And despite his leadership behind the scenes, in early 2010, Mahajan said publicly that he did not believe that a credit bubble was building in Andhra Pradesh.

The metric that Mahajan and his fellow microfinance industry leaders relied upon in the lead-up to the credit crisis fed a false belief that growth could be unending—and that it faced no upper limit or downsides. Despite good intentions to secure the health of their investments and their business, they were deluded by the metric to their eventual detriment.

The myopic focus on ratcheting up loan numbers—like steps in a fitness competition—also created blind spots for the industry's leaders and investors.

Exactly why these metrics were such misleading measures did not become clear to Mahajan until his pilgrimage across India.

When he reached Andhra Pradesh at the end of his journey, Mahajan asked former borrowers what they would do now that the private microfinance lenders had been banned from the state. He talked to a man who bought and sold the secondhand parts of auto rickshaws, and who had stopped repaying his loans. He met with women's self-help groups, and learned about what they had experienced before the crisis.

The groups had a built-in accountability system: Because the collective's credit depended on each woman's diligence in repaying her loans, the women would discourage one another from defaulting. Many would cover payments temporarily if another woman was overextended. The practice offered flexibility, but, Mahajan learned, it also masked each person's ability to pay as a borrower. He heard stories of groups that had shamed and stigmatized women who could not pay on time—and even withheld day care for their children. He saw that many women were not taking out loans to start businesses, but to pay hospital bills for family members or feed their children during a year of lackluster harvest. Generating no income, they took out yet more loans from other agencies to repay prior lenders when they came calling. This pattern could not be sustained even as repayment rates rocketed, no matter how many loans they were offered to cover existing debts.

Mahajan discovered that he had been missing, in a sense, the forest for the trees—not unlike the unfortunate King Croesus. On his pilgrimage, he gained perspective.

Mahajan ended his months of travel in the village of Pochampally in Andhra Pradesh, the place where Bhave had launched the Bhoodan movement sixty years before. The village is also known for its specialty of weaving ikat saris, whose yarns are first tie-dyed to later reveal extraordinary patterns in the woven cloths. Mahajan considered the Bhoodan movement the greatest peaceful land reform movement in history, by which the poor were given millions of acres of land their families had once worked, a rightful inheritance.

After hearing villagers speak and researchers present their findings about the suicides, Mahajan called for a moment of silence. He then told a gathered crowd that microfinance companies should pay the families of those whose suicides had been linked with their loan recovery practices. He rededicated himself to working for the poor, and proposed reforms to his company and industry that he believed could prevent future folly, including deeply investigating each household's ability to pay back loans.

We cannot possibly anticipate or predict everything that might happen to us in our lives. Everyone hopes to avoid catastrophe, and yet some of what befalls us is just bad luck. What we can avoid is steering directly toward the cliff. As we take stock of the meaning of our lives and the progress of our work, we might heed Solon's advice and the lesson of Mahajan's rise and fall. It is tempting to decide that we are doing well or poorly at any given moment, because we have so many ways to measure where we stand today. A monthly or bimonthly ritual of reflection is one way to address this. We can question at frequent junctures what we might be missing, even as we hit near-term targets. Some people I know make monthly lists of what is important but not urgent, posting reminders in prominent places in their homes or offices. Others decide not to place too much stock on hitting or missing a target. It's a way of hedging against the reckless neglect of what matters to them.

We might also find deeper connection to the future by asking ourselves at regular intervals—not just at the end of our lives—what we want our ultimate legacy to be. What will we want to have done when we get to the road trip's end? What do we want to be remembered for, and have we made progress toward that, or have we just amassed more points in the game of life?

BEYOND THE HERE AND NOW

Cracks in the Culture of Instant Gratification

Know when to walk away; know when to run.
—DON SCHLITZ, *"The Gambler,"*
as sung by Kenny Rogers

A myth pervades popular culture that thinking ahead is the province of only certain gifted people. In this picture of human behavior, whether or not people make reckless decisions hinges on their innate abilities. Culture and society are relieved of responsibility, as are businesses and communities.

If you buy into this view, it's easy to feel hopeless. How could you better exercise foresight if everything depends on the fixed stars of human nature?

The misconception traces back to a quintessential experiment testing children's ability to delay gratification: the infamous "marshmallow test," first conducted by psychologist Walter Mischel in the 1960s.

In his now widely known experiments, Mischel gave more than six hundred kids at the Bing Nursery School at Stanford University the choice between immediately gobbling up one of their favorite treats—a marshmallow, cookie, mint, or pretzel stick—or waiting for up to twenty minutes until an adult would return with a bigger payoff of two treats.

Decades later, Mischel discovered that the same kids who delayed gratification at a young age later achieved higher SAT scores and earned advanced degrees at higher rates. They were less likely to struggle with obesity and drug addiction compared with those who couldn't wait very long.

As he published the results, high-profile commenters promoted childhood self-control as the prevailing predictor of life success. "Passing" the marshmallow test by waiting became a theme on *Sesame Street*. A motivational speaker gave a popular TED talk in 2009 called "Don't Eat the Marshmallow!" and schoolchildren wore T-shirts bearing the same slogan. The study inspired stories by major news outlets, many of which confused the correlation between the habit of waiting and later success with the notion that willpower is preprogrammed in just a few chosen ones at birth. The early sorting of kids who passed the test and kids who failed it appealed to parents who craved an immediate answer to the question of how their children would turn out in life.

The studies that have followed the initial marshmallow test, however, tell a deeper story about the importance of environment and culture in how people control their impulses. One research project led by scientists at the University of Rochester, for example, suggests that delaying gratification is not an innate ability, but a choice that some kids make to adapt to their circumstances.

In 2012, cognitive scientists Celeste Kidd, Holly Palmeri, and Richard Aslin tweaked Mischel's marshmallow experiment. They designated one group of children to face reliable conditions and another group to face unreliable conditions. Before the experiment began, kids in the "unreliable" group were promised flashy new stickers and fresh crayons for an art project, but never received them. The other kids got the art supplies they were promised.

The researchers discovered that most kids whose environment delivered reliable delayed results waited longer in expectation of a second treat. The kids who indulged in the immediate reward were those for whom the promise of future treats was worthy of skepticism, or at least highly uncertain. The results suggested that maybe it wasn't the case that some kids were better at self-control than others. Rather, some kids chose to adopt different strategies that suited the conditions and the degree to which they trusted the adult and the promise of the payoff. Circumstances—not merely a child's nature or the amount of information she got about the future reward—made the difference. Maybe what had actually stayed constant in the lives of the kids who performed well in Mischel's marshmallow tests were their life circumstances, not their abilities.

We see this dynamic play out in decisions people make in their daily lives. Someone who does not have the money today to buy a pair of shoes that could last him a few years might choose, because of his circumstances, to buy a cheaper pair that will wear out in three months—even if over time, he ends up paying more than he would if he had bought better shoes. Or he might not think that far ahead because he feels pressed for time, and happens to be passing a discount shoe store. Giving him information about how much money or resources he would save would be unlikely to change his decision. But changing his circumstances, or perception of his circumstances, might.

Culture might be just as influential as circumstances. German psychologist Bettina Lamm, who conducted her own version of the marshmallow test with nearly two hundred children, found that the four-year-old children of subsistence farmers in Cameroon passed the marshmallow test at dramatically higher rates than children in Germany. (The Cameroonian kids also whined less while waiting for the double serving of their favorite treats, and seemed to almost meditate while resisting immediate temptation.) The children from both countries had been tracked since they were three months old, and the 2017 study was the first documented instance of the marshmallow test being given to children outside Western culture.

Lamm does not know exactly why 70 percent of the Cameroonian kids waited for two local doughnuts, known as puff-puffs, while less than 30 percent of the middle-class German kids waited for their preferred treats. Previous studies have similarly shown that less than a third of

American children wait for the second treat. But Lamm and her colleagues did observe that the Cameroonian parents, from the Nso ethnic community, were raising their kids with starkly different customs and expectations. Mothers expected respect from their children, and, in contrast to the German mothers, did not look to their children as often for signals of their needs. The Cameroonian mothers breastfed newborns immediately, before they started to cry, which Lamm saw as preventing babies from needing to express negative emotions as they enter the world. The Cameroonian children, raised in mud-brick homes without electricity on simple diets of mostly corn and beans, were expected to work on family farms and take care of their younger siblings.

The findings raise the prospect that cultural values and practices can drastically change people's relationship to their immediate impulses. Even though the Cameroonian children were much poorer on average than German or American children who have taken the marshmallow test, their culture may have motivated them to value future payoffs.

Yet another marshmallow study, published by researchers at the University of Colorado, Boulder, in 2018, suggests that the norms of a child's peer group strongly affect whether or not the child waits for a second treat. Telling a child he was in a group of peers that shared the same colored T-shirt, and showing him images of other children who waited, seemed to influence what the child chose to do with the first marshmallow.

By contrast, experiments that have tried to link people's biological traits, such as gender or race, to their ability to exercise patience have not proven convincing. There is far stronger evidence that cultural norms of groups shape how people view the future and account for the differences we might see in how people use foresight across different regions, professions, or political parties.

A powerful idea has emerged from this body of research: People can change how they act with respect to the future depending on conditions and cultural norms. The insight is profound, because it offers us the choice to create the right conditions for ourselves and others to practice foresight, rather than just relying on people's willpower.

This idea has long been overlooked by people who prefer to see the marshmallow test as a bellwether that predicts whether a given child will fail or succeed in life. In his final years, Walter Mischel, the test's initial proponent, actively tried to dispel the myth that the ability to defer rewards is ingrained in a person's nature.

Unmasking the myth of the marshmallow test points to a yet more important conclusion: We are not doomed to making reckless decisions. We just have to learn the cultural practices and norms that make thinking ahead possible.

When people try to save money for the future, they often face a dilemma. While saving is good for them financially in the long run, it requires the near-term sacrifice of money available to do whatever they want or need in the present.

I asked Timothy Flacke, who has worked for nearly two decades to help poor and middle-class families in the United States keep afloat economically, whether he knew of ways around that conundrum. Flacke is the executive director of the Boston nonprofit Commonwealth. "People recognize that it's savvy and smart and in their best interest to save," he said. "But it's just really hard."

The mismatch of families' aspirations to their actions can be devastating. When a family is not saving, it finds itself in crisis when needs arise outside a daily or weekly budget. When the Federal Reserve surveyed thousands of Americans in 2015 about their saving habits, it found that 46 percent would not have enough money to cover an emergency expense of just $400.

People in desperation often focus on immediate needs, and for good reason. Sendhil Mullainathan and Eldar Shafir, two behavioral science researchers who have closely studied decision making among poor people around the world, argue that scarcity—of time, attention, or money—focuses people on the present and often leads them to borrow against their own future. It can be tempting to deem this behavior irrational, as it is seen from the arm's-length view of some economists. For people in poverty, however, a focus on the present can be out of necessity the most practical way to make decisions. In their insightful book *Scarcity*, Mullainathan and Shafir explain why the poor take payday loans, even though with high interest rates the loans often dig them deeper into poverty. They write:

> Like all the worthy goals that do not matter when you're speeding to the hospital, the long-term economics of the payday loan do not matter at that moment. This is why payday loans are so attractive— people turn to them when they are . . . putting out a fire. And their best feature is that the loans put out this fire, quickly and effectively. Their worst feature—that the fire will return in the future, possibly enlarged—is obscured.

In one study, Mullainathan and Shafir gave cognitive tests to Indian sugarcane farmers when they were strapped for cash before a harvest, and again when they had more resources immediately after a harvest. They found that the same farmers exercised greater impulse control when they had money. A person who might look reckless when poor could look smart and strategic when flush. Realizing that people who are lacking resources often have a kind of tunnel vision for the present helped me understand why many women involved in India's microfinance crisis went against their own future interest, taking on too many loans and falling deep into debt. It also explains why the poorest families have more trouble heeding hurricane predictions.

Bindu Ananth is the board chair of an organization based in Chennai, India, called Dvara Trust, which provides financial services to India's poor, from investment education and insurance to savings accounts and loans. Dvara's research shows that the poor are highly reliable in paying back formal loans to banks and microfinance companies, and that most are even eager for the structure of a loan repayment schedule. But Ananth believes that loans have too often been relied upon by companies, governments, and aid organizations as silver bullets for the money problems of the poor.

"If you have a health issue, having an expensive loan might work in the near term, but not the long term when you have to repay," she told me. "Too often, the poor, having no other options, are using loans as a substitute for savings or insurance."

The question is: How can you help people save who have so little?

Flacke and his colleagues have noticed one trend that seems at odds with the saving and borrowing habits of the poor, at least in the United States: their high rate of playing the lottery. One survey of Americans shows that 38 percent of those in low-income brackets—and 21 percent of Americans overall—feel winning the lottery is the most practical way to accumulate a large sum of money. They fantasize about jackpots they are unlikely to win that could solve their financial woes. Meanwhile, less than a third of the lowest-income households in the United States regularly save. Those same families account for a disproportionately large share of lottery ticket spending.

Lotteries have an appeal, Flacke learned, that savings accounts—at least those that keep your cash readily available for withdrawal—don't have. They present the prospect of winning big today, rather than waiting for a long-run payoff. Jackpots have an immediate allure, breeding an illusion of the possible. Scratch tickets also offer people the experience of small, incremental rewards—a dollar or two—that make us feel like we are lucky and should play again.

In 2009, the Michigan Credit Union League, working with Flacke and Commonwealth, decided to try using the features of lotteries to encourage people to save for their own future. They offered people who deposited money into their own savings accounts the prospect of winning sweepstake prizes in that month. In addition to earning interest on their accounts, the savers earned entries into a prize fund that operated like a raffle, with frequent drawings and a $100,000 jackpot. The pilot turned out to be wildly successful—more than eleven thousand people deposited more than $8 million into new accounts. This was a surprise, even to those who designed the scheme: The financial crisis had just devastated middle- and low-income families in Michigan, many of whom lost their jobs, and yet they were putting money away for the future. The scheme proved more popular than savings options that had high interest rates.

The idea for the pilot in Michigan was informed by the work of economist Peter Tufano, the founder of Commonwealth and now the dean at Oxford University's Saïd Business School. Tufano has studied the success of Premium Bonds in the United Kingdom, which function as hybrids of lotteries and savings. The British government launched its Premium Bonds program in 1956, to encourage saving after World War II. For the past seven decades, between 22 and 40 percent of UK citizens have held the bonds at any given time. The savers accept lower guaranteed returns than comparable government bonds in exchange for the prospect of winning cash prizes during monthly drawings. Tufano's research shows that people who save under these schemes typically do so not instead of saving elsewhere but instead of gambling.

In contrast to real lotteries, people are never at risk of losing their initial investment when they save in these programs. Often, they forgo only some portion of the potential interest earnings on their savings accounts (or the market's rate of return on bonds) and they stand the chance to win either a big jackpot or small cash prizes. In other words, they are playing the lottery but stand no chance of losing any money. In fact, they are saving for their own future rainy days.

Prize-linked savings programs were illegal in much of the United States prior to 2009 because of restrictions in state lottery and gaming laws. After the Michigan pilot, several states began to make exceptions for lotteries to be held by banks and credit unions in cases where they encourage savings. In 2014, Congress passed a law creating a similar exception at the federal level. Since the Michigan program was launched, more than 80,000 Americans have saved through programs like this in credit unions, and more than 200,000 have used a similar feature on a prepaid bank or store card. A smartphone app called Long Game, designed by a Stanford University researcher in late 2016, also draws on these principles. Long Game allows people to easily open savings accounts—with as little as $60—and add to them with their phones. It enters the savers into a sweepstakes with prizes up to $1 million. Within six months of its launch, the app had more than 12,000 users, who were adding on average about $50 to their accounts each month. Their deposits are backed by the Federal Deposit Insurance Corporation.

More than a dozen U.S. states have seen banks or credit unions offer prize-linked savings programs in recent years, as have more than a dozen countries. Two banks that launched prize-linked savings programs in Argentina in 2007 saw their deposit amounts and number of customers saving money rise by about 20 percent within six months.

One barrier to adopting the programs is that commercial banks don't profit much from savings, so they tend not to invest in innovations like these unless they see the importance of building long-term relationships with new customers. In recent years, low interest rates have put a damper on recruiting banks, Tufano says, because it takes too many savers to be able to offer monthly prizes of $1 million, a figure that captures people's imaginations and encourages them to save more. South Africa's prize-linked savings program, which Tufano also studied, became so successful that the lottery commission felt threatened and the government shut it down. The government wanted to hold a monopoly on lotteries to raise funds to pay for public services such as infrastructure and public education. The long-standing problem with government lotteries, however, is that they exploit the poorest people to pay for public services—like a regressive tax on the neediest members of society. Prize-linked savings reverse this dynamic by getting people who gamble to save for themselves.

What I find clever about prize-linked savings programs is that they wed people's immediate temptation to gamble with their long-run aspirations to save. The underlying idea is that we can motivate ourselves in the present to do something that is good for ourselves in the future. This seems to me the flip side of setting near-term targets—instead, it's finding ways to lure ourselves to stay on the long course.

I once used a tactic like this for a less lofty cause. When a friend of mine from college decided to run a marathon in Houston a few years ago, I wanted to help him finish his first-ever 26.2-mile race. My friend has a whimsical spirit and loves bright colors. During the marathon, I surprised him at key mile markers by showering him with glitter—what's known in certain activist circles as "glitter-bombing." (Some glitter explosions have aggressive intent and are not welcomed by their targets, but luckily, mine were well received.) At the finish line, my friend told me the glitter bombs had given him the boost he needed to endure the most arduous parts of the race.

The strategy is not so different from giving a kid a balloon or a colorful toothbrush at the dentist's office for enduring the discomfort that prevents later cavities. As long as the rewards we pick aren't at odds with our long-run goals, there is evidence that the method can help us work a year of night shifts, learn a difficult language, or go back to school to earn a degree.

What I think of now as the glitter approach can also be useful for helping people prepare for coming disasters. One expert on flood readiness I know has spent a lot of time failing to persuade people in flood-prone regions that they need to protect their homes from the coming water. What works, she says, is sharing with homeowners specific tactics that can reduce their annual flood insurance premiums. People are persuaded to spend time and money putting in rain gutters or sealing basements when they know they have something to gain in that year— something concrete and immediate, not just avoiding the bogeyman of a future flood.

Communities and insurance companies could give people more financial rewards today for what they do to prepare for future climate disasters—especially given the cost to government and private companies of responding to disasters after the fact. At the very least, governments should not penalize people for taking action to prepare for natural hazards. Until 1990, people in California who made home improvements to protect themselves from earthquakes actually saw their real estate taxes rise. Then the state's voters wisely passed a ballot initiative to keep people's tax rates from increasing when they retrofitted homes to meet higher seismic standards.

As an even better glitter tactic, communities might give people tax rebates for preparing for disasters. Some cities and states in the United States, such as Houston and New Jersey, have offered generous buyouts for people to leave their homes after disasters. This could be done in advance in highly hazardous areas, and combined with limiting the number of decision points when people are tempted to favor present concerns. Home mortgages, which are long-term contracts, could be contingent on getting long-term flood or hazard insurance rather than relying on people to make wise choices about insurance each year, giving them the repeated chance to privilege immediate savings over preparing. This is similar to hedge fund investors limiting the number of times they look at their portfolio, and therefore sell stock, in a day.

The glitter approach might even offer a way to curb global warming. For decades, proposals to reduce greenhouse gas emissions from transportation and electricity have faltered because of their failure to address people's drive for immediate wins. Most people and businesses don't want to endure the costs of higher gas and electricity prices today for some abstract idea of a less warm planet in the future. In recent years, a number of groups, both conservative and liberal, have proposed that the United States take a glitter approach to reducing carbon dioxide pollution—by putting a cap on the emissions that would raise the price of fossil fuels like coal and oil, and offering people tax credits or sending them dividend checks quarterly in the mail. The revenue to pay for the dividends could come from taxing polluting fuels like oil and coal. Taking an approach like this could make clean energy policy attractive to more people in the short term, as opposed to a perceived sacrifice.

Glitter explosions can't work for every situation, of course. Sometimes, we find ourselves overwhelmed by immediate temptation, and we can't hold back.

If ever there was a place where recklessness could spread like weeds across an abandoned lot, it is Las Vegas. It's a place where some people directly profit from the shortsightedness of others. To me, Las Vegas today is a dystopic analog of Walt Disney's Epcot Center in the 1980s. It is a model of society's future, assuming it continues on its current trajectory—both a telescope and a mirror for our time. I went there to experience what I imagined to be the most extreme obstacles for heeding the future. And I thought I might also have fun.

Inside a casino, one can learn a lot about why we neglect the future for the sake of our immediate desires. Through careful design choices, casino managers create an environment where many—if not most—people will choose to abandon wisdom about the future for the sake of an alluring fantasy.

It starts with the abundance of temptations. Inside the Planet Hollywood Resort and Casino, I noticed a blind man holding a cocktail in one hand and his cane in the other. He looked no more disoriented than I was as my attention batted around like a pinball. The din was a cacophony: cards shuffling, glasses clinking, an arcade of electronic melodies, the *click-clack* of chips against chips. Waitresses plied gamblers with free glasses of champagne. Neon lights beckoned them to "The Pleasure Pit."

The temptation to try your luck and the anticipation of winning feel familiar but amplified in a casino, as if you were holding ten cell phones, each lighting up with flirtatious messages. You dwell in expectation of the next moment's rewards, not tomorrow's or next week's aspirations. At some point, you might find yourself, as I did, mindlessly handing over your hard-earned cash to a pleasant blackjack dealer named Pam.

Casinos take advantage of the human tendency to get hooked on instant gratification. Neuroscientists have shown that when people get a dose of certain rewards, whether it's chocolate or an orgasm or a flashing gold star on a screen, a surge of dopamine is released in our brains. It's a chemical transmitted between our neurons that trains us to want to unleash more of it—as soon and as often as possible. That's why a winning combination at the slot machine, the ping of an incoming text, a sip of a sugary drink, or a social media "like" often leaves us craving the next thrill.

As we fulfill our immediate desires, we get caught in compulsion loops, relentlessly seeking the next fix. We are left with the constant feeling of having just watched a cliffhanger in a suspenseful TV series. Instant gratification becomes addictive, obscuring our view of delayed consequences and the passage of time itself. This kind of fixation on the present is nothing like the Buddhist ideal of being in the present moment—what Ram Dass called "being here now" in the 1970s or what Jon Kabat-Zinn dubs the mindful state. Rather, being wired for instant gratification manifests as anticipation of the immediate future, a state in which the present state is perpetually dissatisfying.

Anthropologist Natasha Dow Schüll, a professor at NYU, has analyzed the ways that casino managers seize upon people's pursuit of dopamine rushes. They offer free perks like food and drinks that stand in for consistent wins when people are actually losing money—in order to encourage customers to keep placing bets. Video gambling machines, which have been added in droves to casinos since the 1980s, are particularly designed to breed addiction, Schüll contends, giving people the opportunity to experience the rush of a near win on a cycle of mere seconds. These contemporary slot machines manufacture sensory rewards, melodic sounds and colorful displays on their screens. The small rushes we get from these ratchet up our desire to play the game again and again, ushering us into a kind of trance where we are oblivious to our own interests and to how long we've been sitting there. In Las Vegas, you find ceilings of casinos painted like daytime skies, backlit to create the illusion of eternal daylight, even as nocturnal gamblers linger at machines and card tables into the witching hours.

While Las Vegas is an extreme environment nearly synonymous in the imagination with indulgence, the technologies we rely upon in our daily lives are designed to similar effect: They offer abundant temptations—all the world's knowledge at our fingertips, and all the people we know (and many whom we don't) reachable within seconds. They compel us to obsessively seek the next fix with sounds and notifications that signal retweets, likes, and new messages. We carry the casino with us in our pockets every day.

Like salty bar snacks that make beer drinkers crave more beer, the prevalent tools of our time, from instant messages to one-click shopping carts, condition us to getting what we want right away. We come to expect everything to go faster, rapping our fingers when we get caught in a checkout line with a slow-moving cashier. Not only do we measure time more often—we also feel that we have less of it. In a trajectory that dates back to Polaroid cameras and microwaves, ubiquitous consumer products have compressed what we deem an "instant" into ever shorter increments of time.

Ed Finn, the director of the Center for Science and the Imagination at Arizona State University, documented how Google worked to improve its widely used search engine to eliminate mere milliseconds of delay between our questions and the answers yielded in a Web search. The Google "autocomplete" feature that now finishes your queries before you have finished typing them—or in some cases before you have finished thinking them—is aimed at anticipating your desires and gratifying them ever more quickly.

The display of digital time is emblematic of how technologies can erase past and future from our perception, locking our focus on the instant. For generations, analog clock faces and watch dials portrayed the time of day in context, letting us track hour, minute, and second hands moving from past to future and in a cycle bound to repeat itself. Digital clocks show us only the immediate, a moment pregnant with the next moment.

In the seventeenth century, Isaac Newton drew a contrast between absolute, mathematical time and time as we perceive its passing—"apparent" time. Literary scholar Harold Schweizer argues that the compression of apparent time we experience amid technological progress leaves human beings feeling increasingly alienated and anxious. "The indignities of waiting in a culture of the instant," he writes in *On Waiting*, "are also the discomforts of being out of sync with modernity." He points to a reason we could expect our anxiety to worsen in the future. "If acceleration of social change leads to a contraction of the present, in waiting the present is painfully prolonged." He compares shrinking time to the sudden flickering of a camera shutter. "The acceleration of time in modernity, in other words, has greatly accentuated the tediousness of waiting."

In an era when waiting has become intolerable, the pursuit of instant gratification has become almost inescapable. We act so much like addicts when it comes to smartphones, internet searching, and social media that brain scientist Peter Whybrow calls these technologies "electronic cocaine." Even if we don't face the immediate threat of overdose like other addicts, we are still training ourselves to routinely sacrifice our aspirations for the future in order to indulge urgent desires.

We'll likely become yet more saturated with technologies that offer faster and more seamless experiences in the decades to come. And even if we could somehow rid ourselves of these tools entirely, it's hard to imagine that we would want to do so. Digital technologies are the steam engines of our era that power all we make and do, connecting us to each other across distance. Not many people would choose to turn their backs on products that make things quicker and easier, here and now. I do not cling to any Luddite fantasy of escaping this culture entirely. Like many people, however, I want to know how to resist my immediate urges to focus on the future when I need to most. I want to be able to lift my gaze when I am most tempted to veer recklessly off the road.

A cousin of mine is a poker player who regularly visits casinos in Las Vegas and other cities. Unlike most gamblers, however, he usually walks out of a casino after winning a lot of money. He once told me about the professional poker players he knows who earn six and seven figures a year playing the game strategically, not recklessly. These pros go to casinos all the time, and some even live in Las Vegas. They are not victims of gratification culture, but its masters.

When I visited Las Vegas in 2015, I watched a poker tournament with a $1.5 million prize pool in a room upstairs from Planet Hollywood's main casino. The hall doubled as a dull conference room with dingy carpeting, stiff black chairs, and tepid overhead lights. Bottled water and cans of Red Bull replaced swish martinis handed out by cocktail waitresses just a floor below. A muffled voice over a loudspeaker broadcast each new round, sounding as bored as the announcer at a high school swim meet. A janitor rolled a miniature dumpster past the tables to collect sandwich wrappers as the tournament marched on. The scene, with hundreds of poker players in baseball hats sitting around dozens of tables, felt more like a bingo game in a nursing home than an elite gambling event on the Vegas Strip. It might have been more entertaining to watch a tooth decay.

Professional poker is not the bastion of people who intrinsically have greater willpower than the rest of the population. It is rather a subculture that has emerged—with its own norms, its own language, and its own habitats—within the casino and within Las Vegas. Like punk rock in the 1980s, the subculture is a kind of counterculture, whose proponents are adamant about articulating what distinguishes it from the mainstream culture of gambling that they scorn. They see themselves as hackers of Vegas, strategists unlike the careless, naive tourists.

Poker stands apart from other casino games, not just physically, but because it is possible to play it for a living. The odds are not stacked impossibly against a player with skill and some good luck. Most of the players who endure in poker get the opposite of instant gratification, the pros say. They earn their livelihoods not in single dramatic plays, but over time in many games. Successful players persist from hand to hand, level to level, within tournaments that last for days. They hold on to their chips and fold their hands frequently, waiting until a good opportunity arises to make a big bet. Winning players have to withstand strings of losing hands without retreating in order to make it to higher levels in a tournament. Weak players, by contrast, struggle to avoid the urge to try to play and win each hand and what's known as "vendetta poker"— trying to out-bet other people on the basis of petty disagreements or battles of ego.

The lingo of poker pros reinforces their values. The long path from learning the game to making a living is called "grinding" and is viewed with a kind of reverence for putting in the time to perfect one's craft—as if it were an apprenticeship in a Renaissance sculptor's studio. When a player wastes her earnings at poker on playing craps or blackjack on the casino floor, or on expensive booze or nights out, it's called a "leak," to connote that the boat might eventually sink. The language shores up the norms of the subculture, to which not all conform but many aspire.

Enduring loss is a badge of honor in professional poker culture. Dutch Boyd, a pro known for his success in the 2003 World Series of Poker, put it to me succinctly: "A lot of people think they want to be poker stars, but actually sitting down to get good at it kind of sucks." Boyd's experience had been marked by years of sustained losing, punctuated by wins that put him in the black overall.

Another player, Ronnie Bardah, grew up getting dropped off at video arcades while his father gambled away the family savings at racetracks. Bardah trained himself over the course of nearly a decade, starting out at low-stakes poker tables to gain experience playing amateurs (known in pro poker as fish) before he entered the high-level tournaments and cash games he now plays around the world. He took home his first six-figure prize purse at a tournament in 2010, seven years after he started playing poker seriously. Like many other successful poker players I met, he had made short-term sacrifices for the sake of greater but uncertain long-term rewards. He had resisted temptations to get more, like a kid passing the marshmallow test. And he escaped sharing the fate of his father, whose gambling had left him impoverished.

A pair of British interviewers talked to several poker players in the United Kingdom in 2008 and 2009 and found a pattern that reinforced what I discovered in my encounters in Las Vegas. Successful professional poker players, those who earned upward of $150,000 in profits each year, differed from those who played recreationally in that they did not try to immediately make up for losses, taking a broader perspective. They seemed to have figured out how to detach emotionally from their losses in the moment.

The poker players' attitude toward loss reminded me of something sociologist Marshall Ganz told me about why certain social movements endured over time, while others failed. Ganz said that a critical factor is how movement leaders characterize setbacks—whether they describe them as learning moments that will make the movement stronger or as failures that signal that the adversary is too powerful. In other words, seeing losses from a bird's-eye view might prevent people from abandoning a cause in particularly hard moments. Case Western Reserve University political scientist Karen Beckwith has studied what she calls "narratives of defeat." Tracking labor movements in Great Britain and the United States, she documented that a positive framing of loss helped some movements survive where others failed. Martin Luther King, Jr., said that "we must accept finite disappointment, but never lose infinite hope."

Professional poker players inhabit their own environment, one that partially shields them from the temptations of the rest of the casino. Poker tournaments—even in Las Vegas—typically take place in rooms that stand apart from the main floor of a casino, like the hall I saw at Planet Hollywood. This is calculated on the part of the casino managers, because their businesses do not make as much money from poker players—who take one another's chips rather than lose them to the house—as they do from the other gamblers. (The house keeps a small percentage of the pots from cash games and some of the entry fees from tournaments.)

Poker players tend to linger, making slow gambles rather than pouring their cash steadily into slot machines or losing it rashly in large sums at the roulette wheel. The casino prefers to offer its perks to those outside the subculture rather than those within it. The players, in turn, benefit from the neglect; the relatively spartan habitat has few of the casino's features that entice people to indulge their immediate urges. No women exposing deep cleavage, no free top-shelf liquor, no bells or lights to flatter you for every coin dropped. The compulsion loop of seeking the next fix is fractured.

The existence of poker culture as an island within an environment designed to encourage shortsightedness raises the prospect of cultivating countercultural habits more broadly, to resist instant gratification in our lives.

Author and entrepreneur William Powers initiated such a practice within his family a few years ago, inspired by the teachings of Plato, Seneca, and Thoreau, who each in his time sought refuge from a world becoming more connected and hectic. Powers created a habitat within his family home each week to break compulsion loops. Every weekend, the family observes a secular "Internet Sabbath," abstaining from Google searches and e-mailing, and avoiding their digital devices. In doing so, they create a temporal island that allows for deeper thinking and more focused social interactions. Author Pico Iyer has also advocated creating time periods within contemporary life to relish stillness and quiet.

Movements are afoot in corporate America to create spaces where people put away their digital technologies for periods of time. Former Walmart executive and author Neil Pasricha proposes scheduling an "untouchable day" every week—where you have no meetings, no phone calls, and don't check your e-mail or social media feeds. He credits his own adoption of the practice with leaps in productivity and success completing multiple long-term projects.

Since time is elastic in our perception, how much we feel we have of it—and hence how impatient we feel—can be influenced by our environment. Fast music cues people to be less patient, it seems, than slower rhythms. In a set of experiments, UCLA business professor Cassie Mogilner found that people who write letters to sick children, or take time on a Saturday morning to help a friend, are afterward more likely to feel like they have time to spare in their lives than people given a trivial time-wasting task, or those simply invited to relax.

Culture makes a difference, as it likely did for the Cameroonian children who waited for their treats longer than German kids did. In Singapore, I routinely witnessed people backing their cars up into parking spaces in lots and garages, a cultural quirk that one person described to me as being "ready to leave in a rush if ever necessary." There are forward-thinking habits we might similarly adopt as norms within families and neighborhoods. While people have long suggested that there are biological differences that make Singaporeans or Chinese people more oriented toward the future, my inquiry has led me to believe that cultural and subcultural practices matter far more. The Chinese stock market, where investors can behave in highly speculative and short-term ways, for example, has different cultural norms from those of the autocratic Chinese government, which has oriented itself toward empire building over centuries.

Most poignant is that a subculture can redefine people's expectations and their sense of urgency. At a Quaker school in New York City that I visited, the day begins with a meditative all-school gathering in a sparse hall. Students and teachers can speak when they feel moved to speak, but the cultural norm of the school and in the Quaker tradition is that time and silence should elapse before anyone responds to another person's remarks. The quiet can last for minutes that feel eternal to the uninitiated. This is a form of countercultural practice, to wait to reply—or not reply at all—in an era when we increasingly feel the urge to respond immediately to every message we receive. To watch teenagers in that habitat, waiting in silence, contemplating, listening, thinking, is to witness a small revolution.

In a previous era, places of worship and civic halls may have served as gathering places where people consistently came together to think past their immediate concerns and focus on the future, and even the afterlife. Today, for the most part, such spaces remain at the margins, set apart from the places where work and life happen rather than the habitats where we dwell. We might go to yoga classes or meditation centers for an hour a day, or to the dinner parties of certain friends who will give us blistering looks if we pull out our smartphones. Then we return to the virtual casino of our daily lives. We need to cultivate more environments that make it possible for us to think ahead—and especially, I believe, secular spaces where people without identical belief systems can practice patience together. These might take form as parks or courtyards, or rooms in homes and workplaces where slow conversation and quiet reflection are encouraged. In the meantime, it would help to know what rituals we can take up on our own.

Playing poker is like planning for the future in one sense: You have to make decisions with only some information, and a lot depends on chance. You can control what you do only with the cards and chips you have; you can't control what cards you or your opponents are dealt, or what the other players do. You make choices in the moment amid uncertainty about how it will all pan out.

In another sense, however, playing poker is not like making decisions in the real world. In poker, there is a large but fairly defined range of possible outcomes—combinations of cards, moves made by other players— many of which can be anticipated. Many seasoned professional players even know, roughly, the odds of different outcomes. For the situations we face in which we *can* anticipate what we might encounter—the predictable obstacles that get in our way—the tricks of successful poker players can be illuminating.

Matt Matros, a World Series of Poker champion who has earned more than $2 million playing professional poker, shared with me one of the rituals he credits for his success.

When he first started playing poker in the early aughts, Matros loved the thrill of winning. But because he also hated losing, he almost never won a tournament. He was too timid to play hands that were worth betting on, and so he did not earn enough chips to go head-to-head with more aggressive players in the late stages of tournaments. Trying to avoid momentary loss made his prospects for winning dim in the long run—in a similar way to how some investors lose their cool when they see dips in their stocks.

Matros, a self-described geek, earned a degree in math in college. Cerebral players like him started flocking to the pro poker scene after a twenty-seven-year-old accountant from Tennessee with the apt name Chris Moneymaker won the World Series of Poker Main Event in 2003. Moneymaker was a dark horse; he had spent years as an amateur learning and practicing poker, mostly online. His first in-person tournament was the 2003 World Series of Poker, in which he turned his $39 buy-in for a qualifying game into a $2.5 million victory.

What made Matt Matros a poker champion was coming up with a distinctively nerdy game plan. He forged an overall strategy before he ever got to the table. He would aim for a certain ratio of bluffs to "value bets"—betting when he actually had a good hand—in a given game. When he didn't have a good hand, he would fold a certain percentage of the time and bluff the other times.

The strategy required that he anticipate the scenarios he might face at the table, so he had a plan for how to react regardless of his passing fears of losing a hand. But he didn't have to creatively concoct scenarios like the Wellesley psychologist Tracy Gleason before her camping trip. Because this was poker, he could calculate the chances of winning or losing he'd have with different combinations of cards, something that's not always so easy in life. Whenever Matros was in an intense moment, where the exact card pattern was not one he had already anticipated, he emulated the ritual he had practiced before of thinking through options. It was as if he had rehearsed the scene of a play and could think back to his scripted lines, with some improvisation.

I wondered if Matros's techniques would work only for the über-cerebral, elite thinker. Then I came upon the work of Peter Gollwitzer, a professor of experimental psychology at NYU. In the 1980s, while leading a research group at the Max Planck Institute in Germany, Gollwitzer began studying how people can adhere to their long-term goals amid immediate temptations. He found that most people he encountered were not lacking motivation to set such goals. The trouble came when they tried to keep on course when confronted with their detrimental short-term urges. He soon started testing a technique that resembles Matt Matros's strategy for playing poker, which he dubs implementation intentions, or if/then tactics. In hundreds of studies spanning all kinds of contexts—from eating better to completing homework assignments to saving money to avoiding the impulse to react to people on the basis of their skin color—Gollwitzer and his colleagues have demonstrated the power of people taking the time to anticipate in advance the obstacles they might face when meeting future aspirations. For example, people who want to eat healthier might jot down all of the possible temptations they'll face in a week to eat junk food, and then set up a plan to respond to each of those urges.

What's surprising about Gollwitzer's research on if/then techniques is that it reveals that the more difficult the long-term goal, the greater the power of the tactic. It works better, in other words, for the challenges that elude people's sheer willpower. He has also shown that for groups of people who have the most difficulty with impulse control, patience, and perseverance, the tactic works even better. Schizophrenics, alcoholics, and children with attention-deficit/hyperactivity disorder, when guided to use if/then tactics, resist distractions and temptations and defer gratification at high rates in his studies. They get more out of the tactics than those considered normal. Matros's poker habit is a secular ritual we may be able to use more widely.

In practice, adopting an if/then ritual is fairly simple—even obvious. In fact, I hesitated to write about it for that reason, until I learned about its surprising, untapped potential beyond the realm of trivial decisions. Suppose you've decided you don't want to check your e-mail tomorrow morning so that you can work on a long-term project. To prevent yourself from rashly ditching your aspirations in a moment of weakness, you could think through situations you might face that would tempt you or distract you to make a different decision. Then you could come up with ways to respond. For example, you might decide the following: "If I realize I need to respond to an e-mail, I'll jot down a reminder in a notebook with bullet points so I don't forget to send it later." Or: "If I feel tempted to check my e-mail because I hit a hard spot in my project, I'll get up from my computer and take a short walk outside instead." The more concrete the action you plan, the better the technique works. Mental images and positive actions—stating what you *will* do as opposed to what you *won't* do—are more powerful.

If/then rituals ask people to envision themselves taking a desired action in the future, not just to anticipate future scenarios. This struck me as a way of addressing what psychologist Tracy Gleason pointed out about the anxiety some people experience when imagining the future. With an if/then tactic, people picture themselves with agency in a scenario. This is a way of turning dread into a plan for the future—and guiding your future vulnerable self in a calmer or aspirational present moment. When people make and state their specific plans to vote on Election Day, for example, they are far more likely to show up at the polls. The ritual's use is constrained, however, to the scenarios we can practically envision in advance. (Much of what we face in the future is unpredictable, which I will explore in later chapters.)

What intrigues me most about if/then rituals is their potential to stop reckless decisions that affect not just our own lives—but the lives of others.

In 2009, Peter Gollwitzer and his colleague Saaid Mendoza made a stunning discovery through experiments in which they asked people to play a computer game they call the Shooter Task. In the game, a series of images of men appeared, some holding objects and some holding guns. The objective of the game was to shoot the men who appeared with guns and not shoot those who held other objects such as wallets or phones. (The choice involved pressing a labeled key that read either "Shoot" or "Don't Shoot.") The players were asked to work fast, and make reflexive decisions in the moment. They had half a second before their time was up for each of the eighty images.

The game showed images of both black men and white men. Previous studies with this game had shown that people "shoot" the black men who are unarmed at far higher rates and more quickly than they shoot unarmed white men. They are also more likely to *not* shoot armed white men than armed blacks. The 2009 study's participants—none of whom was black—when first given the game to play, exhibited this same pattern of what's known as "implicit bias." In most cases, people who show these tendencies are not people who would consider themselves racist or who would consciously mistreat others because of their gender, race, age, or appearance. But in situations where they must act quickly, hidden biases can often dictate people's actions and even override their intentions.

Mendoza and Gollwitzer gave half of the study participants an if/then tactic before they played the game. These players were warned not to let other characteristics of the target, aside from what they were carrying, dictate their response. The researchers recommended adopting the following strategy: *If I see a person, I will ignore his race!* The participants were asked to repeat this strategy in their minds three times, and then to type it into a box after it disappeared from the screen. Those who used the strategy had far fewer mistakes—they "shot" more armed people of both races and let the unarmed go free.

Public school teachers and principals around the United States are starting to take up this research and try to use it to prevent rash decisions based on race. Kathleen Ellwood, a school principal in Portland, Oregon, discovered in 2012 that in her K–8 school, nearly 90 percent of the students being sent to the principal's office were black—even though they represented only 15 percent of the student body. National trends echo that pattern. In 2018, the U.S. Government Accountability Office analyzed data across K–12 public schools and found black students are being suspended and expelled at far higher rates than white students regardless of a school's poverty level and the type of discipline. Black middle school students have almost four times the chances of being sent to the principal's office to be disciplined as white students, despite having no demonstrably higher rate of misbehavior.

Ellwood, who is white, began her teaching career at a public school in Long Beach, California, during the violent uprisings and racialized conflicts that took place there in the early 1990s. She found that even the most hardened teenagers—members of gangs and kids who had seen people murdered in front of them—could be brought to tears thinking of times when a teacher or principal told them at a young age that they were stupid or would not amount to anything. In her view, the experiences the children had early on in school shaped their future dreams and life choices.

When public school teachers in the United States discipline students of color, and particularly young black children, studies show, they do so more harshly than they discipline white students for the same behaviors. The result is that more students of color miss valuable time in the classroom because they are sent to the principal's office, and given detentions and suspensions. This affects their achievement not just in school but in life—feeding what's known as the "school-to-prison pipeline." The kids who miss school are much more likely to land in the criminal justice system.

It's not the case that all the teachers who send kids to the principal's office are consciously racist or have a desire to punish certain students—in fact, research shows that most have no such intentions. But in exasperated moments, impulsive decisions reflecting ingrained biases become more likely. Teachers, like all of us, are exposed to portrayals in the media and popular culture of black people as criminals, and those images shape unconscious views and actions.

University of Oregon professors Kent McIntosh and Erik Girvan call these moments of discipline in schools "vulnerable decision points." They track discipline incidents in schools around the country and analyze the data to show school administrators and teachers that such moments are often predictable. When teachers are fatigued at the end of a school day or week, or hungry after skipping lunch for meetings, they are more likely to make rash decisions. Some moments are more specific to a given classroom or teacher, but can still be anticipated—for example, a kid with whom a teacher had a power struggle interrupts in class.

This bears out the link Eldar Shafir and Sendhil Mullainathan have shown between scarcity—in this case, of time and attention—and reckless decision making. It's similar to the pattern that hamstrings the poor from saving for their future.

Teachers and principals betray their own intentions in moments like this. Even Ellwood, who has spent more than twenty years trying to address racial disparities in education, told me that she has at times found herself treating students of color differently from how she would like.

Ellwood invited McIntosh to train teachers in her school to adopt their own rituals to guide how they respond to students in vulnerable moments. McIntosh asked the teachers to make an advanced plan for how they would respond in a future moment when they would likely be tempted to issue harsh discipline—and to state the plan clearly and concretely, aloud and on paper. The idea was to create an impulse buffer—a delay between the moment of frustration and the decision to discipline.

For example, suppose a teacher recognized that she got frustrated on days when she skipped lunch, and especially when kids interrupted her as she was speaking. She might make an advanced plan, using this technique, and state it aloud as follows: *If a student speaks out of turn, I will drop a pencil and take a deep breath before deciding what to do.* Or she might choose to say the following: *I will take three steps back with my hands behind me if a certain student acts up.* McIntosh has trained hundreds of teachers around the country to use such tactics.

Ellwood found that after her school's teachers—and also students and administrators—adopted the if/then rituals, the skewed disciplining of black students dropped significantly, and fewer students were sent to the principal's office overall. McIntosh and Girvan are studying the use of such tactics across multiple schools in the United States to see if they consistently reduce discrepancies in discipline, and to see how the practices might help reshape school culture. The work is promising, but it remains to be proven whether it works across schools and classrooms.

Simple rituals like these that we can each adopt have the potential to help us heed the future. But I fear they will never be enough for situations where factors other than our personal urges prevent us from thinking ahead. A CEO is unlikely to make an investment in research and development at her company if her shareholders and board would punish that choice by firing her. A mayor will probably not block development on a fragile coastline if real estate companies pledge to campaign against him in the next election. A farmer is hard pressed to take better care of his land if the crops that sell better are those that erode it. To make it possible to think ahead in such situations, we have to look beyond the individual.

PART TWO

Businesses and Organizations

4

THE QUICK FIX

Cues to the Consequences

Always carry a flagon of whiskey in
case of snakebite, and furthermore,
always carry a small snake.
—W. C. FIELDS

A mistake is not a singular moment. A constellation of circumstances, rather, shape a momentary decision that goes awry.

Most of us can rattle off stories of shady organizations that have made mistakes. We remember the headlines from scandals and the faces of fraudsters who actively courted future disaster for the sake of immediate profit. A car company gets fined billions of dollars after designing a way to cheat on emissions tests. A bank falls under investigation for creating millions of fraudulent checking accounts. A charity loses donors because it overlooked its field workers' payments to prostitutes.

The unwitting ways that organizations encourage reckless decisions may pose an even greater threat, however, than the cheating we find so repulsive. The work of John Graham at the National Bureau of Economic Research puts eye-popping scandals into perspective. He has shown that more money is lost for shareholders of corporations—including those of us who have 401(k)s or pensions managed by institutional investors—by the routine, legal habit of executives making bad long-term decisions to boost near-term profits than what is siphoned off by corporate fraud. In isolation, these are decisions that seem innocuous, but they add up to a pattern of recklessness.

Many of the most damaging mistakes that organizations make are preventable. If businesses, nonprofits, and government agencies can learn new tactics, they can seize future opportunities and heed warning signs of disaster. They are not doomed to a dismal fate.

Discovering this has dampened my cynicism about the potential for businesses to exercise foresight. By studying how organizations err and triumph, I have come to see their untapped power to both envision the road ahead and avoid steering off course. They, like each of us, have a choice.

For most of her career, Dr. Sara Cosgrove has been leading a battle against superbugs—the deadly bacteria spreading around the world that are resistant to known antibiotics.

Cosgrove is an infectious disease physician, a medical school professor, and the head of what's known as the antibiotic stewardship program at the Johns Hopkins Hospital in Baltimore. She is one of the world's leading experts on infectious diseases resistant to treatment, and is called upon to consult governments and businesses around the world. She is understated, warm, and prone to finding the humor in sticky situations. Her dark eyes light up with resolve when she talks about the crisis she has been trying to quell for decades.

Cosgrove wanted to be a doctor from the time she was a child. She was frequently sick with colds and taken to the doctor's office, where she would open up drawers filled with mesmerizing cotton balls and glass tubes while waiting to be examined. At home, she mixed "medicine" bottles of hydrogen peroxide, shampoo, and bathtub powder and pretended to treat her kid brother—luckily, he never took a dose. While in high school in the late '80s, she was captivated by a story in *Rolling Stone* magazine about the AIDS epidemic that was killing thousands of Americans each year.

As a med student, Cosgrove thought she'd spend her life treating HIV and AIDS patients. In medical school, she interviewed women with HIV about their plans after death for their children. During her fellowship training in 2001, however, she became concerned about a different, overlooked problem. She noticed that patients in the hospital who contracted drug-resistant infections were dying at higher rates than other patients. Many of the deaths could be prevented, she thought, if only doctors could make better decisions about how they treated patients. Cosgrove got inspired to tackle the problem, and it has consumed her ever since.

"It's a puzzle to solve," she says. "Each bug has its own story." What Cosgrove has discovered fighting superbugs holds broader lessons for how organizations can avoid reckless decisions.

Superbugs are on the rise worldwide today, killing more than 700,000 people each year. Gonorrhea that won't yield to antibiotics is spreading throughout the world, and drug-resistant infections like *E. coli* are killing people in hospital wards and nursing homes. Contagious tuberculosis now infects people in more than 120 countries. By the year 2050, more than 10 million people are expected to die each year from superbugs.

For hospitals, this is a liability—and a nightmare. When an outbreak spreads through a hospital, patients expecting to be healed run the risk of instead contracting a deadly disease. The most vulnerable patients—those with cancer or receiving organ transplants—are at the highest risk of getting infected. Superbug outbreaks in hospitals erode people's trust in doctors and the health care system.

The rise of superbugs is not a surprise to people like Cosgrove who study infectious disease. It was both foreseeable, and foreseen—and the problem is largely avoidable. Today most medical professionals—doctors, nurses, patients, and hospital managers—even understand *how* to prevent it. The challenge is that the culprit is also a cure.

Antibiotic drugs first came into widespread use on a cold November night in 1942.

A conga line snaked across the dance floor that evening at the Cocoanut Grove, a frothy speakeasy-turned-nightclub that lit up the insides of a drab warehouse in Boston. It was the liveliest place to party in town. A crowd of a thousand swarmed the club after the Boston College–Holy Cross football game. Celebrities, mob bosses, drunks, soldiers, and sports fans transported themselves to the tropics amid the Grove's zebra-striped sofas and fake palm trees.

A week before, the composer Irving Berlin had performed at the club to promote his forthcoming movie *This Is the Army*, which featured Ronald Reagan and the smash-hit song "God Bless America." Across the Atlantic, Hitler had just invaded Vichy France. The Grove's retractable roof opened to the stars on warm evenings, but on that chilly night, blue satin canopies draped the ceiling, and a single revolving door served as both entrance and exit.

Eyewitnesses reported this sequence of events: A sixteen-year-old barboy struggled in a dark corridor to screw in a lightbulb that furtive lovers had loosened for privacy. To see the socket, he lit a match, which he dropped on the floor. Soon, the nightclub's fake palms and drapes erupted into flames and people rushed the narrow exit to escape. Fifty years later, the Boston fire department concluded that methyl chloride, a flammable gas, had been circulating throughout the club, leaking from the defunct air-conditioning system.

Nearly five hundred people died in the Grove fire, and hundreds of burn victims poured into the city's hospital wards that night. Their open wounds would be vulnerable to staph infections, a death sentence for fire victims in those days. It was the worst fire in Boston's history and one of the deadliest in the country's history, killing hundreds more than the Great Chicago Fire of 1871 that burned for two days across miles of the city.

There was a glimmer of good news, however. A company in New Jersey got word of the fire and rushed a truck northward to Massachusetts General Hospital, where many of the victims were being treated that night. It carried in its cargo an obscure drug that had been taken by fewer than a hundred people in the United States thus far. It was called penicillin, and doctors treating the Cocoanut Grove fire victims credited it with a high rate of patient survival, and with making it safer to create skin grafts for the most severe wounds.

The success of the drug amid the Cocoanut Grove tragedy spurred the U.S. government to back companies making antibiotics. Penicillin-spiked drinks, mouthwashes, and soap soon flooded the marketplace. "Thanks to PENICILLIN . . . He Will Come Home!" boasted one company's ad in *Life* magazine in 1944, plugging the drug as the great healing agent of World War II. Until the mid-1950s, Americans who wanted penicillin didn't even need a doctor's prescription.

In the early twentieth century, before the spread of antibiotics, premature death touched every household. People died easily and routinely from pneumonia, diarrhea, and scarlet fever. Newborn babies lost their mothers to septic shock. In some cities, infections killed 30 percent of children before their first birthday. In 1927, an Austrian doctor won the Nobel Prize in Medicine for treating the dementia of syphilis patients by infecting them with malaria—a strikingly bad idea, though it was praised in the desperate era before antibiotics.

With the spread of penicillin and other antibiotics, it became far less scary for people to get an insect bite or give birth. Getting a cough was no longer a likely death sentence. Antibiotics were nothing short of miracle drugs, making it possible to have an organ transplant or triple bypass surgery without dying from an infection. They also made sex safer: American soldiers in Vietnam who frequented brothels received regular penicillin shots to ward off gonorrhea. Some prostitutes got them, too, courtesy of the U.S. military.

Everybody loves a quick fix for a nasty problem, and antibiotics have long played the part. But today they have also extended themselves beyond the realm of cure into the province of reflexive precaution. Farmers feed them to healthy chickens, just in case. Doctors dole out prescriptions for common viruses that our bodies could fight off with time and rest.

"Because antibiotics came on the scene and saved so many lives, we lived with the idea that antibiotics were not harmful from the mid-1940s to the 2000s," Sara Cosgrove told me. "The question for doctors was not when to give them out, but why wouldn't you give them out?" The fact that antibiotics, the twentieth-century cure-all, could cause harm has suffered the handicap of being counterintuitive: How could something so miraculous actually pose any danger?

Today a majority of the antibiotics prescribed in the United States and the United Kingdom for coughs and colds—and likely far more in China and the developing world—are not the right treatments for patients' illnesses. The unneeded antibiotics pose downstream danger by killing off microbial communities in our bodies that help us stay healthy, while also leaving us vulnerable to deadly infections.

These dangers were understood long ago by the doctor who started the antibiotics revolution. Alexander Fleming, who discovered penicillin's powerful ability to kill bacteria in 1928, warned presciently in 1945 that, as time passed, the overuse of antibiotics would spur bugs to evolve to become resistant to the drugs. For a few decades, however, such long-run consequences were masked by new drugs that replaced the old.

Superbugs might be less of a concern if drug companies could just keep making new antibiotics for doctors to usher in once the old ones become inert, like changing the engine oil in a car. But most of the easy-to-develop antibiotics have already been discovered, and it is expensive and difficult to invent new ones. Drug companies can make far more profit developing a statin or painkiller that patients take for years of their lives than they can from antibiotics that patients take for several days. It's a losing battle, with an offensive surge in superbugs, and a dwindling supply of drugs for defense. Antibiotics have become a scarce resource that people are using up too fast without regard to future consequences, like the freshwater in an aquifer or the coal in a mine.

Sara Cosgrove wants to keep the remaining arsenal of antibiotics we have potent. To do that, she says with a laugh, "we have to stop prescribing antibiotics when it's stupid."

Doctors who prescribe unneeded antibiotics, however, don't tend to do so out of ignorance. They often do so betraying their own good intentions.

Early in her tenure at Johns Hopkins, Cosgrove and her colleagues conducted a study to see whether what doctors knew matched what they actually did in practice. By surveying and observing residents, she found that young doctors who said they would not prescribe antibiotics under hypothetical scenarios in reality prescribed them in those very situations. Aspirations to value their patients' long-run health and preserve the potency of antibiotics for the community were not matched by their actions in the moment. Nor did doctors' knowledge about the "right" decision correspond to making better decisions. If these were if/then rituals, they had failed.

It's hard to imagine why doctors, steeped in scientific knowledge and among society's most rigorously trained professionals, would prescribe antibiotics that people do not need—especially given the consequences for public health and even for their individual patients. But when you think about the conditions doctors face as they prescribe, a clearer picture emerges. In the moments doctors are in front of patients, urgency often prevails over foresight. The cues they get encourage their neglect of the future.

Fear drives hospital doctors to prescribe antibiotics even when they are not needed, Cosgrove has found. Doctors are afraid of what might happen to a patient—and to their own ability to practice medicine—if they don't use every tool at their disposal.

With inaction, a hospital patient could get sicker, or even die. Doctors envision potential lawsuits. They are haunted by cases where a patient fares badly, and often fixate on what they could have done differently. Rarely do they experience—or even know—when they have done too much for a patient, whether it's ordering extra blood tests or exams, or offering medication that causes a downstream problem. Meanwhile, other factors discourage waiting to see what makes the most sense.

The culture of contemporary medicine places an emphasis on giving something to patients. In a sense, having a treatment option creates the default impulse to use it, so a doctor can't be seen as withholding a potential cure from a patient. While we are fortunate today to have thousands of technologies and drugs to diagnose and treat disease, from MRIs to chemotherapy, they also present dilemmas. With so many tools, it's easy to lose sight of the basic question of whether and when it makes sense to treat a patient at all. This also drives up health care spending.

National directives have encouraged antibiotic overprescription in hospitals rather than correcting it. One policy in the 1990s took a toll. To address high death rates from pneumonia, the U.S. government's Centers for Medicare and Medicaid Services, the largest single entity that pays for health care in the country, stipulated that when patients with pneumonia symptoms come to a hospital, they should be treated with antibiotics within the first few hours of arrival. Intended to help pneumonia patients, the initiative encouraged immediate—and often incorrect—prescribing of antibiotics before a definitive diagnosis could be made. Researchers began to document *Clostridium difficile*—a superbug that inflames the colon, often contracted after "good bacteria" are killed off by prolonged use of antibiotics—in patients treated for pneumonia with antibiotics. They found that a large share of those patients were likely to have never had pneumonia in the first place.

At doctors' offices, patients are typically not as sick as those who end up at the hospital. Yet doctors and nurse-practitioners in primary care report feeling a lot of pressure to prescribe antibiotics—often coming directly from their patients. For some doctors, the trust that patients place in them creates a sense of obligation. A dad hands the surgeon his newborn baby, an only child that took a decade to conceive. A teenage athlete comes in with a chest cold that she has to kick before a tournament that could determine her college scholarship. It's hard for doctors to look past the person in front of them to consider some vague future threat, says medical sociologist Julia Szymczak, who has spent years researching the factors that shape how doctors use antibiotics. (While some drugs are also marketed heavily to physicians by pharmaceutical companies, most antibiotics today do not fall into that category.)

As doctors and nurses rush between patients, the demands on their attention mount. Long shifts lead to fatigue and backlogs, leaving them yet more pressed for time. In a study of more than twenty thousand visits to primary care clinics, researchers found that the likelihood that a doctor or nurse-practitioner would prescribe an unnecessary antibiotic increased significantly as her shift wore on, as she became more likely to fall behind on her schedule and to experience fatigue. As one doctor I know put it, "It takes five times longer *not* to prescribe an antibiotic, because then you have to explain to the patient why you are not prescribing it."

It's not just in medicine, of course, that fatigue, time constraints, and social pressure push people to make reckless decisions. Recall the work of behavioral scientists Sendhil Mullainathan and Eldar Shafir. They have shown that scarcity of time is like scarcity of money, in that it drives people to focus on immediate needs at the expense of the future. They point to the example of NASA's Mars Orbiter, which crashed in 1999 as a result of a miscalculation in programming the firing of its reverse thrusters, which were supposed to slow the craft down so it could enter the Mars orbit. Even before the crash, NASA admitted that engineers working long hours to meet a tight and immovable deadline to launch the orbiter were making mistakes. The error that caused the crash was a simple failure to convert between the metric measure of force, newtons, and the English measure, pounds.

What workplace today does not place people under considerable time pressure, or hand over important decisions to people who are overworked and tired? At the very least, every organization in our era is overburdened by information flow. Tight deadlines, long shifts, and stress are hallmarks of organizations that ship freight, develop mobile apps, fly passenger airplanes, police neighborhoods, broadcast the news, serve meals on weekend nights, build rockets, rescue fire victims, fight and settle lawsuits, make self-driving cars, and teach children. This trend should be cause for concern. The decisions people make in these organizations affect all of us, and too often their focus by necessity is on the urgent and immediate, and not on the ultimate consequences.

Social scientists Roger Bohn and Ramachandran Jaikumar called this kind of decision making in organizations "firefighting"—a pattern of suppressing urgent fires without attention to what might still be smoldering. In a 2000 paper, they pointed out that organizations in constant crisis neglect what is important over time. They create this culture, in part, by overburdening workers and rewarding firefighting instead of the prevention of fires.

In both hospitals and primary care offices, doctors and nurse-practitioners prescribing drugs tend to be far removed from the consequences of their decisions. They might not see what happens to their patients down the road. Immediate factors hold more sway than the hypothetical infection that a patient might get, or the potential superbug that might emerge globally over time. We see this dynamic not just with antibiotics, but also in the reflexive prescribing of painkillers by doctors and dentists. This has contributed to the opioid addiction crisis in the United States, with rising death rates.

The worst dangers of overprescribing tend to be spread across populations, while individual doctors treat one patient at a time. This doesn't help matters. The decisions that doctors make to prescribe drugs hinge on the pressures of that moment and the people involved in that interaction—whereas the consequences of superbugs are borne by the collective, whether a hospital, a community, or a society. Myopia is rewarded by a quick, positive exchange with a patient, whereas foresight on behalf of the collective goes unnoticed or even punished if it is done at the cost of just one pneumonia patient. Old-fashioned economists call such situations where individual motivations are not aligned with the collective good "tragedies of the commons." (I turn to this dilemma in more detail in the next chapter.) The trick is for organizations to find ways to align individual people's decisions with the greater good they want to achieve.

In the early 1990s, when Sara Cosgrove was a medical student at Baylor College of Medicine in Houston, an outbreak of an antibiotic-resistant strain of *Acinetobacter* was spreading in the intensive care unit of Ben Taub Hospital. The largest of three county hospitals in Houston, Ben Taub was also the teaching hospital for Baylor, which means its doctors were professors at the medical school and its students learned how to practice medicine in the hospital's wards.

Hospital staff tried to stop the outbreak by isolating the patients who contracted the superbug, and they launched a campaign to encourage doctors and nurses to wash their hands. But the outbreak kept spreading. Patients were suffering, and some dying, from an illness they contracted at the hospital itself.

The hospital managers decided to enforce a new requirement. Doctors who wanted to prescribe any of six antibiotics first had to call and get approval from an infectious disease specialist. On staff twenty-four hours a day, the specialists would gauge whether the case merited an antibiotic before the hospital pharmacy would dispense it.

Researchers tracked what happened for possible pitfalls. Would patients die because of the delay in getting drugs? Would some patients fail to get the antibiotics they needed?

What the researchers found was that with prior approval, antibiotic use and resistance in the hospital dropped. Bacterial infections in the hospital became dramatically more susceptible to the restricted antibiotics. More patients affected by the Acinetobacter outbreak survived. Meanwhile, survival rates stayed the same for patients in the hospital overall. That means the extra step did not deprive sick patients of drugs they needed—patients who actually needed antibiotics still got them within twenty-four hours. And because the scheme reduced overall prescription rates, it saved patients and insurers money typically spent on antibiotics.

Elsewhere, in teaching hospitals across the country, requiring doctors to seek approval before prescribing antibiotics yielded similar results. It reduced costs and fought back superbugs. In short, a buffer created by managers kept doctors from succumbing to the urge to immediately prescribe a drug to a patient. There was now a delay between the moment of diagnosis and the moment of prescription, and a third party was involved who had a long-term view of the patient's health and a broader view of the concerns of the hospital and the community. In some cases, prior approval bought time for doing a bacterial culture to determine whether a patient had an infection that would respond to a given antibiotic—or if he had a virus best treated another way.

There was, however, a major flaw with the scheme: Doctors didn't like it. Many felt policed, and others suspected that hospital administrators just wanted to save money and were not concerned with patients' health. Some doctors engaged in what's known as "stealth dosing"— prescribing more of the antibiotics not on the restricted list or waiting until the specialists went home for the night in the hospitals where they were not on duty twenty-four hours a day. "Doctors got sick of it," Cosgrove says. "They would say, 'Leave me alone! Why are you telling me how to treat my patient?'"

At Johns Hopkins Hospital, Cosgrove has since put into place a program that improves upon prior approval, and complements it. She works with a team of seven infectious disease experts—all whip-smart women who are doctors and pharmacists, whom she trains and mentors. The team looks closely at the studies of different drugs and at their costs to determine which should be carried in the hospital pharmacy (known as a formulary) and which have to be requested by special order—which builds in a delay of a few days.

Cosgrove's core team has also trained a bigger team of thirty specialists, including pharmacists, infectious disease fellows, and doctors, who go around the hospital to talk to doctors who are requesting restricted antibiotics. The staff are on call not just to tell doctors what they can't do, but to help guide their decisions. In the process, they also create a small delay that allows for more reflection and data-gathering. I followed Cosgrove and her colleague one afternoon as they made rounds in the hospital's medical intensive care unit. They had a rapport with doctors and physician assistants, and asked non-accusatory questions. They helped doctors figure out what was happening with mysteriously sick patients, and when to taper down doses of antibiotics.

I see Cosgrove's team as akin to the sailors whom the mythical Odysseus asks to tie him to the mast. They stuff beeswax in their ears so that they will not succumb to the song of the Sirens and wreck their ship. Although Odysseus is tempted by the song, the sailors refuse to untie him. Organizations often need dedicated teams like those sailors and the team at Johns Hopkins to prevent reckless decisions, to play the role of tying the hands of those vulnerable to immediate pressures. At Johns Hopkins, the scheme does not work perfectly all the time—doctors sometimes resist having their decisions questioned or get angry when they are told they cannot prescribe a drug. And yet the hospital far surpasses others like it in its prevention of reckless prescribing, and many doctors have even begun to see their track record as a source of pride.

Most antibiotics are prescribed to people who show up with coughs and colds at their doctor's office—not to hospital patients. While hospitals are on the front lines of fighting superbugs in their wards, they don't directly influence the majority of bad prescriptions that give rise to the bugs.

Outside hospitals, the scheme that worked at Johns Hopkins is not practical. Regular doctors' offices don't typically have infectious disease specialists, nor the resources to hire teams to advise them.

Most efforts to curb antibiotic prescribing at doctors' offices in recent decades have had little to no success. The tactics have ranged from teaching doctors about the risks of superbugs, giving them financial bonuses to reward better prescribing, and planting pop-up reminders in electronic health record systems about the downsides of doling out too many antibiotics. The flawed assumption underlying these tactics has been that doctors simply need more information about future risks—or to have more money on the line—to make the right decisions. Too little attention has been paid to what makes good judgment about the future—foresight—possible.

In recent years, however, a new way to stop bad prescribing in doctors' offices has been showing promise.

Daniella Meeker, a researcher at the University of Southern California, has discovered three tactics that reduce bad prescribing in doctors' offices. In 2011, Meeker launched a study of forty-seven primary care practices in Los Angeles and Boston. Over the course of a year and a half, she tracked nearly seventeen thousand patient visits for which antibiotics were not the right choice, because the patients had either the flu or another kind of viral infection. She made various interventions and tested the results. In 2014, she launched a separate study across five primary care practices, trying another intervention.

The first technique that worked in Meeker's studies was as follows: Anytime a doctor prescribed an antibiotic, she received a prompt in the electronic health system when she typed in the order, asking for a justification of the prescription. The doctor or nurse-practitioner could choose not to explain, but in such cases they would be told that "no justification given" would be entered in the patient's permanent medical record. (The system also let the doctors cancel if they decided not to prescribe the antibiotic after all.) The prompt served as an impulse buffer without taking away the doctor's autonomy to make the decision to prescribe. It also created an emotional deterrent based on her professional image and the potential for shame and loss: the fear of an entry in the patient record that might reveal to both the patient and the doctor's peers that the decision had been wrong. It was an automated version of the stewardship team at the Johns Hopkins Hospital that advised doctors—and tied them to the mast during pivotal moments.

The second tactic that worked was sending a monthly e-mail to doctors telling them whether they were top performers—or not—in prescribing antibiotics under the right circumstances. Like a report card calibrated to reflect social norms, the e-mails allowed doctors to privately see where they stood relative to their high-performing peers. This was not abstract information about future risks, but data given in context—customized to the person making a decision and held up against the choices made by others with better track records. Some utility companies use a similar approach, sending households monthly reports comparing their energy use with that of their more efficient neighbors. Overall, the tactic reduces people's use of electricity. Telling hotel patrons that the norm is to reuse towels has a similar effect. Research suggests that organizations can use simple techniques to make people feel that they are part of a cultural group with shared expectations to use foresight. This reminds me of the kids in peer groups with norms that encouraged waiting for the second marshmallow.

Meeker and her colleagues found that these two interventions had greater effect than giving doctors information on antibiotic stewardship or simply observing their actions. The two approaches also had a far greater influence than did suggesting alternatives to doctors via pop-up screens. What worked was setting new cultural norms and buffering their impulses, not just information.

The third successful technique that Meeker documented was asking physicians to mount posters in their exam rooms. In her 2014 study, posters in English and Spanish bearing doctors' photographs and signatures expressed a commitment to prescribing antibiotics only when needed and shared with patients the risks of taking antibiotics. After three months, bad prescribing dropped in the practices with the posters relative to those without them. What I find interesting about this technique is that doctors communicated to their patients—shaping not just their own professional culture but the broader culture of medical encounters in order to tame social pressure in the moment. People who make public commitments to donate money to causes, vote in elections, or recycle similarly follow through more, research shows, than those who just have a private intention to do such things.

In 2014, the British government undertook a national experiment with the report card approach—sending letters to thousands of doctors in England who were prescribing the most antibiotics per capita in their regions. The letters, from high-profile British leaders, let those doctors know that they were prescribing more antibiotics than 80 percent of their local peers and suggested alternatives to writing a prescription in the heat of the moment, such as giving patients advice to care for themselves while sick. Researchers from the UK government's Behavioural Insights Team found that these letters corresponded with a substantial decline in the rate of antibiotic prescription, with an estimated seventy thousand fewer antibiotics given to patients in a six-month period. The letters had cost very little, but they had saved significant sums of money spent on medicine by the national health care system and protected public health.

California has since tried a similar approach to prevent inappropriate opioid prescribing, creating a state registry of physicians who prescribe the drugs and patients who request them. The state's medical board as of 2016 requires all doctors to track when they prescribe opioid painkillers and why. Doctors can see if a patient has been shopping around for opioids, and medical board and law enforcement investigators can identify doctors who are prescribing the drugs recklessly. By 2018, many other states had emulated California, creating their own digital opioid drug registries—though some restrict access to the data more than others.

Some of these state drug databases make patient data accessible to the police. This raises legitimate concerns about privacy, especially given the way our legal system criminalizes addiction. As of this writing, federal courts have ruled it constitutional for medical and law enforcement professionals to access prescribing databases without a warrant. In my view, tracking doctors' prescribing patterns is a good idea, because they are professionals serving the public. But patient data ought to be kept private.

What if, instead of banning certain behavior or appealing to reason, more organizations could set conditions that spur people to exercise foresight? The prospect of doing so, it turns out, has intrigued thinkers dating at least as far back as the eighteenth century.

Benjamin Franklin, for one, thought of himself as an arbiter of virtue, and he freelanced in ideas for influencing human behavior. As one of the founders of the United States and the preeminent American inventor of his time, he concerned himself with posterity. Upon his death in 1790, he bequeathed a total of £2,000 sterling—the equivalent of giving about $100,000 today—as gifts to the cities of Philadelphia and Boston and their respective states, in a trust. He designated a portion to be spent one hundred years after his death—and the remainder to be spent another hundred years after that. By the time two centuries had passed, his legacy gifts were worth about $6.5 million, and the two cities and states began to use the money for public causes. Philadelphia has given grants to high school students learning crafts or trades, and Boston built a technical college. Franklin popularized the phrase "Time is money," and his actions embodied it.

Franklin professed to take the long view of business transactions, not quibbling over petty issues in order to preserve relationships and the potential for collaboration in the future. He made a conscious decision, for example, when he became postmaster in Philadelphia, not to seek retribution against a competitor newspaper owner who had refused to carry Franklin's newspapers when he had previously held the post.

Franklin was not prone to rash behavior, except on rare occasions. He once dunked his argumentative and erudite childhood friend John Collins—an alcoholic, by Franklin's telling—into the Delaware River after Collins refused to take his turn at rowing their boat.

Franklin's lifelong quest for moral perfection borders on being downright annoying, at least for those of us who can't claim to be anywhere near perfect. In his autobiography, addressed to his estranged son and to future generations, Franklin extolled thirteen virtues, ranging from temperance to sincerity to humility, and kept a scorecard of his own actions so he could develop each virtue into a habit. He believed the motivation of one's public image and social norms could shape behavior. For example, he confessed that he wasn't actually humble, but he knew appearing to be so would earn him the admiration of others. Against his inclinations, he practiced patience in conversations where he disagreed with people, first pointing out instances where their point of view could be seen as right. Eventually, the practice became habit. Franklin attributed his influence in public councils not to eloquence but to this perceived humility. The public persona, he felt, had along the way made him the person. Franklin's reasoning sheds light on why exam-room posters and report cards have worked to prevent doctors from making shortsighted decisions about drug prescribing. They, too, had public personas and social norms motivating them.

By the time Franklin was serving as an American delegate in Paris, he'd grown a bit lax on some of his virtues and had taken up the Parisian habit of staying out late and sleeping in. In a satirical letter to the editor of the *Journal of Paris* in 1784, he estimated that 64 million pounds of candles could be saved in six months if all Parisians went to bed as it got dark and woke at dawn. (Another of his espoused thirteen virtues was frugality.)

Though Franklin exaggerated about the laziness of the French, his thrift and inventive spirit inspired him to propose a scheme for persuading people to live by the sun's light. His most impassioned suggestion was to create a cue in the environment to cajole people into keeping daylight hours. "Every morning, as soon as the sun rises, let all the bells in every church be set ringing; and if that is not sufficient? Let cannon be fired in every street to wake the sluggards effectually, and make them open their eyes to see their true interest."

Daylight saving time did not become an official practice in Europe or the United States until World War I—well after Franklin's death. The idea was to save electricity in the summer by giving us another hour of sunlight in the evening. But more than a century earlier, Franklin saw a way to shift people's perceptions of time so that they could make decisions for their self-interest. Marking time differently by setting the clocks ahead an hour, or by waking people with clamoring church bells, was intended to cue people to save energy and resources for the long run.

For a long time, daylight saving time was a boon for urban families and businesses who saved money and energy with the extra hours of evening sunlight. The original purpose of observing daylight saving time has since been rendered moot, however, by modern habits. People who now crank up their air conditioners while they are awake and the sun is still shining, or turn on the lights in the dark morning hours before they go to work, do not save much electricity or money because of daylight saving time. And many parents today find it hellish to adjust their young children's bedtimes to the twice-a-year ritual of resetting the clock.

Regardless of the current wisdom of advancing the clock, changing the environment to influence behavior remains a powerful idea. It's a way of designing culture, and Franklin ordained himself an architect.

A full century after the death of Benjamin Franklin—and before the start of World War I—another thinker emerged who invested far more time testing ways to encourage foresight by design. She was also the first woman to earn a medical degree in Italy, at a time when women were discouraged from walking alone outdoors. As a medical student at the University of Rome in the 1890s, Maria Montessori dissected corpses alone after dark because it was considered indecent for a woman to be among both naked dead bodies and fellow male medical students. She took up smoking cigarettes to mask the smell of the cadavers. One professor remembered her as the only student to show up to classes during a historic blizzard.

Montessori's early years as a doctor in the psychiatric clinic of Rome brought her to insane asylums, where children deemed mentally ill or slow languished in barren rooms. She became an expert in childhood maladies and was called to the slums of Rome's San Lorenzo quarter, where poor children spent their days without toys, books, or teachers. Through her research and observation of children labeled as deviants, Montessori came to believe that most kids could thrive if their environments were designed to allow them the freedom to make their own choices, within some constraints. At the time, the idea was radical: The prevailing belief was that intelligence was hereditary and immutable in children.

One day early in her medical career at the dawn of the twentieth century, Maria Montessori was struck by the mental focus of a malnourished three-year-old girl. She watched the toddler as she spontaneously played with a toy that consisted of a wooden block with sockets for corresponding cylinders. The girl did not get distracted by the kids singing and dancing around her as she explored and examined the object. To test her concentration and resolve, Montessori picked up the armchair the girl was sitting on and placed her and it on a table. But the child kept going, repeating the exercise of putting cylinders in the block—at Montessori's count, forty-two times. When she finally stopped, Montessori thought that the girl looked as if she had come out of a dream, energized and not at all exhausted by the repetitive task.

Most schools, Montessori observed, immobilized students "like butterflies mounted on pins" and attempted to motivate them with prizes and punishments instead of their own interests. In the early 1900s, Montessori left medicine to begin forging her educational philosophy. It focused on the "prepared environment"—classrooms designed to free children to explore, but also guide them toward decisions that help them learn and develop a work ethic. When she arrived in the United States in 1913, she was celebrated in New York headlines as the most interesting woman in Europe, one who had revolutionized education and done what was then unimaginable, teaching mentally ill and disabled people to read and write. Eventually, she applied the method to so-called normal children and spread the idea that school should start at age three and that classrooms should be designed to captivate the innate curiosity of children.

My stereotype of a Montessori student once consisted of a free-range kid roaming aimlessly through a schoolyard. In 2006, however, I actually met several Montessori students. I noticed their patience and focus, and how they absorbed themselves with tasks. I later learned that the far-thinking founders of Google, Sergey Brin and Larry Page, and of Amazon, Jeff Bezos, had all studied in Montessori schools. It could have been a coincidence, of course, but I began to wonder if there might be something under the hood of a Montessori education that teaches people to delay gratification and persist over time, in the manner it takes to invent groundbreaking technologies or build a successful company.

Hal Gregersen, the executive director of the MIT Leadership Center, confirmed my suspicions with actual research. In interviews with five hundred well-established entrepreneurs and inventors, he found that about a third attributed their innovative ability to the early support of either Montessori teachers, or teachers in what Gregersen deemed Montessori-like schools.

The distinctive design of the Montessori environment seems to matter. Most conventional classrooms strive to have as many objects of a kind as there are students, in order to avoid conflicts and allow everyone to do the same task at once under a teacher's supervision. Contemporary Montessori classrooms, by contrast, particularly for three- to six-year-olds, have single objects of a kind. At a given moment, only one kid can typically examine or play with an abacus or a tower of blocks. By limiting the objects of a kind, Montessori classrooms build in the need for and the practice of delaying gratification: A child won't always get the object she wants to learn about or play with right away, but she will eventually. Teachers encourage kids who covet a particular object to exercise executive function, creating a plan for what they'll do next and how and when they might return to the object later that day or the following day. The plan gives them the ability to exercise patience in the moment. This is similar to how the if/then rituals helped poker players and high school teachers.

You rarely, if ever, hear teachers in a Montessori classroom say no—instead, they'll ask kids what they would like to choose as an alternative to stealing blocks from another kid or sticking a finger in an electrical socket. When two or more kids reach for the same object at once, teachers gently guide a child to choose another activity. When conflicts arise, which they inevitably do, mediation happens at a "peace" table or corner, where a child holds an object such as a stone when talking, and passes it over to the other child when listening. The tools build in a slow rhythm for pacing a conversation that is prone to breaking out into an argument. Kids who want to get a teacher's attention place a gentle hand on his arm, with the notion that the physical connection creates the possibility to overcome immediate urges and be more patient.

"Montessori teachers are masters of redirection," Martha Torrence, president of the board of the Montessori Schools of Massachusetts, told me. Torrence is also the head of Summit Montessori, a private-school community of nearly a hundred kids set in a gabled Victorian mansion in Framingham, Massachusetts, built four years before Maria Montessori was born. The building, painted in pastel yellow, teal, and mulberry, feels more like a storybook gingerbread house than a school. From the halls, kids' voices and the pattering of their feet can be heard, but it's nothing like the utter chaos I pictured when I thought about a school of kids left to make their own decisions—nor anything like what I experienced in my own public school education in Ohio. It was an elite enclave, but perhaps there was something to it that could be adapted more widely.

As I observed teachers in a Montessori classroom directing children away from objects of temptation in the hands of other kids, I could not help recalling my time living in Cuba when I was a college student at the University of Havana more than fifteen years ago. I was captivated by what happened at the bus stops, where people often had to wait for up to two hours for the arrival of an overpacked blue-and-white diesel-spewing bus, the unreliable conveyors of the public through the city. The Cubans had developed a system to manage inevitably long waits in line in a flagging economy—not just for public transit, but also for food rations, movies, and ice cream.

At first, I failed to see that people were waiting at all. People gathered at a bus stop not in any apparent order, but scattered in clumps, attending to gossip, reading the newspaper under the shade of an almond tree, or buying paper cones filled with peanuts from passing vendors. I soon learned that the scheme for waiting involved asking, when you arrived at a bus stop, *¿Quién es el último?* (Who's the last one?) Someone would raise a finger and say, *Soy yo.* Then everyone would go back to their conversations or daydreaming until the bus arrived, at which point they all fell into a perfectly ordered line without jockeying or conflict—because each person already knew whom she followed in line, and to get behind that person.

The elegance of the system the Cubans designed for waiting was that it allowed people to distract themselves from the urgency they felt waiting for a bus, or the temptation they felt craving their ice cream cones on sweltering tropical afternoons. Rather than waiting single file, with body and attention oriented toward the object of desire, they could pursue whatever they wanted, wandering around a wider area, as they waited. But they also had a plan, like the Montessori students, for securing a place in line to get their due. They designed, by community consensus, an environment that made inevitable and uncontrollable delays in gratification tolerable. It was a shared ritual aimed at redirection from the moment of impulse.

Richard Larson, a data systems professor at MIT, has shown in his research that people wait in lines—at places like the bank or a grocery store—far more patiently when they have distractions and a sense of how long they will have to wait. Savvy companies have taken advantage of such insights. Disney World, for example, keeps its visitors happy in long lines with colorful murals, giant animatronic toys, and clearly communicated estimates of wait times. Think of it as the opposite of designing a slot machine alley at a casino—an environment that makes it possible to wait to get one's object of desire.

When doctors prescribing antibiotics are asked to justify their decisions, or sent letters telling them how they size up to their peers, they are being cued to consider future consequences and to personally feel the stakes of collective dangers. The "report card" letters and exam room posters, in the spirit of Benjamin Franklin, use social norms and public image to drive behavior. The pop-up requests, like Maria Montessori's classrooms, redirect attention and create delays during moments prone to rashness. These methods to prevent superbugs were inspired by the growing field of behavioral economics, which offers a research base to bolster what Franklin and Montessori proposed on the basis of their intuition and astute observation.

Economists Richard Thaler and Cass Sunstein call such environmental cues "nudges." A nudge maker seeks to carefully design the structure that surrounds a choice, constraining the options and tipping the psychological scales in favor of certain kinds of decisions—while still allowing people the freedom to choose. For example, some workplaces make employee contributions to retirement accounts the default choice, requiring people to opt out rather than opt in, to encourage them to save for the future. Some programs automatically trigger an increase in the amount that employees save as their salaries rise, so that they do not feel they are ever losing money in the immediate term. This "choice architecture," in the parlance of Thaler and Sunstein, has been shown to dramatically increase savings rates.

I see techniques like this as a form of subtle manipulation. Organizations can draw on them to cue people to weigh the future heavily for their own good, but also might use them for less innocent ends, like keeping people placated as they wait for something, like a city bus that ought to be coming faster than every two hours. Tristan Harris, who formerly served as a "design ethicist" at Google, writes about how contemporary technology companies manipulate us in this fashion. In some instances, they limit our menu of choices to, for example, the restaurants listed on Yelp (instead of all of those that exist in a neighborhood). Retailers and advertisers often make us opt out, rather than choose to opt in, to their frequent e-mail blasts. We tend to choose default options, and decide from what we are presented with, rather than the full range of possible choices.

If organizations are transparent about why they are designing certain choice architecture, however, the schemes can be carried out ethically. Daniella Meeker and the UK government have shown they can also have great success. In the early 2000s, a similar approach was taken up in the Black Forest of Germany by a utility company called EnergieDienst. It offered several communities renewable sources of energy including solar and wind as the default choice, even though they were more expensive than conventional electricity. A 2008 study showed that 90 percent of the residents had chosen the green power sources, even though they were paying more for their electricity in the short run. Making it the default choice made it more cumbersome to elect the cheaper option, and also signaled that the social norm was to choose the cleaner energy, which was better for the community (and presumably, also the business) over time.

Cosgrove's and Meeker's work shows that with dedicated teams or the aid of technology, organizations can interrupt impulsive decision making and motivate thinking ahead by recalibrating social norms.

Organizations can also design environments to prevent recklessness by reducing stress and time constraints wherever possible. Mullainathan and Shafir advocate creating "slack," offering the example of a Missouri hospital that left one of its operating rooms routinely unscheduled to allow for emergency surgeries, which surprisingly helped doctors do more surgeries overall because they were not overburdened by unexpected overtime and the rescheduling of preplanned surgeries to accommodate emergencies. Bohn and Jaikumar proposed bringing on "temporary problem solvers" who could take on some of the work when there are too many crises for an organization, and limiting the number of problems that can become backlogged each day—whether it's engineering problems in a manufacturing plant or waiting patients in a hospital. They also recommend rewarding managers who manage long-term problems more than those who react to crises, and creating leeway to miss deadlines whenever possible.

In recent years, Cosgrove and her team at the Johns Hopkins Hospital have begun to focus on what happens not just before decisions are made to prescribe drugs, but afterward as well. They track hospital patients after they have received antibiotics and flag what they call "bug-drug mismatches"—cases where there has been a prescription of an antibiotic but the bacterial culture did not justify it or where a patient needed a drug earlier and did not get it. They visit with doctors in the hospital for a kind of postgame rehash, sharing information about what eventually happened to the individual patients. With this tactic, they educate doctors not with general pointers about antibiotic prescription, but with specific stories of patients they treated personally. They bring the abstract, delayed outcomes into more immediate, concrete form.

Cosgrove's research team recently showed in a study of more than fifteen hundred patients that this method of playing Monday-morning quarterback and reconstructing past events had a more lasting effect on inappropriate prescribing than did restricting access through prior approval. The drawback is that it requires a lot of time and legwork, and even hospitals with antibiotic stewardship teams lack the resources to do this across an entire hospital year-round.

What I noticed when I followed members of Cosgrove's team around is that the postgame method does not restrict the choices that doctors have, or redirect or shame doctors during vulnerable moments. It instead draws on real stories to bring salience to future consequences and to reshape the cultural norms of the hospital.

This offers yet another way organizations and businesses can orient people toward the future—by investigating and communicating how decisions play out over time. Telling stories of how past decisions turned out can help people better imagine the future—acting as a bridge between the roads traveled and what lies on the road ahead.

A BIRD'S-EYE VIEW

What Counts in the Long Run

What a pity if we do not live this short time
according to the laws of the long time.
—HENRY DAVID THOREAU,
letter to Harrison Blake

Each year while I was growing up, my parents marked my height on a closet door. It gave me a satisfying sense of progress: I was getting taller. Watching the steady uptick of a key number is reassuring to groups, too. That's why organizations track their numbers of workers trained, meals served, profits earned, tests passed, and offenders punished. The measures seem like a simple, objective way to judge the progress of a program or of a person—all you have to know is whether the chosen number is rising or falling. Philanthropies, investment firms, nonprofit organizations, government agencies, and, of course, corporations rely heavily on such metrics to make decisions.

Meeting their numeric targets, however, is not the same as fulfilling their actual goals. Thousands of people might get trained as farmers or software programmers, but perhaps only a few stay in their new jobs. Hundreds of meals can be served in a neighborhood without ever addressing the reasons people are hungry. Profits can be earned while a company heads toward a blowup, and tests can be passed by students who learn little of use for their lives. More people can be put in jail as crime and violence in a city rise.

In the lead-up to India's microfinance crisis in 2010, it wasn't just individual people like Vijay Mahajan who were deluded by myopic metrics. Businesses and investment firms organized themselves to meet numeric targets, including raising loan numbers in particular regions. The microlending businesses reported to their investors the high loan repayment rates and growing loan portfolios. During the crisis, many microfunders failed, and others fell deep into debt, because they neglected to detect the disaster brewing in their industry.

When we look at such incidents, it may be tempting to conclude that organizations should stop relying on numeric targets. Yet it's possible that things would get worse without them. Agencies might rely on blind faith and unbridled patience, instead of good judgment, to deem whether products or aid programs are really working. Companies might summon the excuse of long-term thinking to justify mediocrity or failure. Leaders might ignore signs of future catastrophe revealed by interim measures. There's a danger of not measuring enough.

In an ideal world, businesses would count what matters for the future, not just what meets their present need to show results. Metrics would help people in organizations see when they are making progress, correct course when things are going wrong, and break up long projects into smaller steps. Data points would be viewed over time, not just in snapshots, to reveal trends like growing markets, steadily rising temperatures, or failing investment portfolios.

But how do organizations choose the right measures at the right time, and avoid myopic ones?

Ravenel and Beth Curry boarded the plane to Oregon with a sense of dread. It was 1998, and the stock market was ablaze with thriving new technology companies. The internet had changed everything, or so it seemed, and Wall Street investors were trading shares of "dot-com" companies—online fashion stores, pet supply merchants, sports news peddlers—at extraordinary multiples, before the companies had shown they could sustain their success. The high prices were contagious, spreading through the market, and, for a year and counting, kept soaring. Tech stocks served as a source of easy money for speculators, who saw the opportunity to keep turning over their shares for more profit.

The Currys had not been cashing in on any of it. Rather, the investment company they had started after raising three kids seemed on the brink of falling apart. Their clients, including wealthy families, pension funds, and university endowments, were calling meetings to pull their money out of the fund. The Currys could not recruit new ones. "Nobody was giving us money, that's for sure," Ravenel told me when we met several years later in New York City.

It was hard for Ravenel to blame those people. For four consecutive years, the fund had missed hitting the baseline benchmark used to size up investors in the near term. It had underperformed relative to the S&P 500 stock index, giving less in returns to its clients than if they had simply invested in a passive fund pegged to the market. Most sane people see little point in paying high fees to a money manager unless they earn *more* in returns than an index fund. And 1998 was shaping up to be the Currys' worst year yet.

They had started their modest fund, Eagle Capital Management, ten years earlier. Ravenel Curry, born and raised in a small town in South Carolina, had spent his career up to that point working for others, in financial firms including J.P. Morgan and H.C. Wainwright and later managing the Duke Endowment. He felt that trading stakes in companies on near-term cycles was not that interesting. As hedge funds sprang up and more traders entered the stock market, deploying ever more sophisticated algorithms for optimizing trades, it was also tough to stand out.

"It's hard to get a competitive advantage in predicting quarterly profits, because it's hard to beat the thousands of people doing it," Ravenel said, leaning back in an armchair in his office in midtown Manhattan. "Plus, it's mind-numbingly boring."

In a world where nearly everyone was becoming a trader, he wanted to be an investor who defied the market consensus. He wanted to buy stock and hold on to it until a company made good on its long-term bet to create new markets or invent new products. Ravenel believed he could not easily put that philosophy into practice while working for someone else; he had to start his own firm. His wife, Beth, who had begun a career in finance after a previous chapter as a homemaker and stay-at-home mom, joined him.

In the early years of the firm, the Currys did not worry about losing clients who did not share their philosophy and style. One client, the former owner of a chain of drugstores in the Midwest, gave a significant portion of his money to Eagle to manage after selling his stores to a national brand. Ravenel told the man that the firm made investments carefully up front based on a company's potential growth over five to seven years, avoiding reactions along the way to noise in the stock market.

Just two weeks after the Currys started investing the drugstore magnate's money, the man called to ask how the investment was going. It was too soon to tell, Ravenel replied.

The man called back every couple of weeks to ask how his investment was faring. Sometimes, he'd have looked at how a stock was trading that week and would call to express his concerns. Each time the answer was the same: It was too soon to tell.

Ravenel realized that the man's background as a drugstore owner had trained him to track progress on two-week cycles. The pharmacy was located in the back of a drugstore, in order to encourage people to make impulse purchases as they walked past the shelves to pick up prescriptions. Drugstore managers would measure shelf space to see what inventory was moving on a two-week basis. In that business, twice-monthly turnover was an indicator of sales strength and, indeed, how the business was doing. "He could not handle the long-run perspective of measuring progress, so pretty soon I told him he could find other firms that would manage his money that way, and he left," Ravenel recalled with a shrug.

In the late '90s, however, the climate was different. Losing a client seemed like it could be the death knell of the investment firm. It was also hard to weigh the consequences of the decision to stay out of the dot-com boom. It had been a difficult year for the Currys and their fund. People who had once believed in their investment approach saw their friends cashing in on speculation in the stock market, trading stakes in companies with highly inflated prices on daily or weekly cycles. As they took their money elsewhere to be managed, they would say that everything in the market had changed because of technology, and Eagle was missing the boat. With so many people making so much money, it was hard not to want to get in on the action—not just for Eagle Capital's clients, but also for the Currys. They were tempted to please their clients by changing course.

Ravenel Curry feared that the upcoming meeting in Oregon, with a client who represented a university endowment, was going to bring more bad news. On the plane, he remembered his reason for starting the firm in the first place. He didn't believe in buying stock at overhyped prices that reflected market consensus at a given moment. He believed in buying companies that were undervalued by the market and had the prospect of growing over time. He saw the stocks of most technology companies during that time as dumb bets, and the market as excessive. But, like his clients, he was growing weary of watching everyone else get rich from them.

Peter Rothschild directed investments for the University of Oregon's foundation at that time, chairing the committee that oversaw its endowment. The foundation had given several million dollars to Eagle to invest in the 1990s. Rothschild liked that it was a small investment firm, because he believed that the university would get more attention as a client. He also shared the philosophy of investing in companies because you believed in their value, not just because their stock price could rise as a result of mania in the market. The move had paid off in the early '90s, as Eagle brought forth significant returns. But now, there was cause for concern.

Rothschild invited the Currys to fly out to meet in person. When they arrived, he pointed out the severe underperformance of the firm. "Are you going to do something about it?" he asked.

In that moment, Ravenel thought he would lose yet another account. It would have been devastating. He decided to override his fear, however, and instead he doubled down. He didn't believe that the bubble in the market would last, he said. They would not be changing anything about how they managed the university's money, or anyone else's, for that matter.

To the Currys' surprise, Rothschild then announced the endowment would be giving them more money to manage. His goal, he told me on the phone, had not been to lambast the failings of the fund, but rather to gauge whether the Currys were keeping their cool amid the madness in the market.

Within eighteen months, the market crashed. Some dot-com companies that had been infused with hundreds of millions of dollars evaporated; others saw their stock values dive precipitously. Eagle fared well and way outperformed the plummeting markets in 1999 and 2000. In just those two years, the gains more than made up for the losses of the previous five. Today, the company has grown to manage more than $25 billion in assets and, on average, earned an annual return of more than 13 percent on its investments between 1998 and 2018. That's more than double the annual return from the S&P 500 during that time. Looking ahead made for a rosier future.

People try to game metrics. Organizations that use numeric targets to drive behavior will inevitably find this happening and their goals undermined. That's the upshot of Goodhart's law, named for the economist who declared it in 1975. Charles Goodhart advised the Bank of England in setting monetary policy from the 1970s until the turn of the twenty-first century.

What this means in practice is that when an organization tells people to hit a metric and rewards them if they do, people will sacrifice other meaningful progress—or even cheat—for the sake of meeting that target. In the nineteenth century, Chinese peasants, paid by European paleontologists for every fragment of dinosaur fossil they dug up, soon started smashing the bones to boost their compensation. The practice bears some resemblance to day trading.

After 2001, public schools across the United States adopted student standardized test scores as their key metric for evaluating school progress and teacher performance. But the numeric target in many instances undermined the true goal of education: student learning. David Deming of the Harvard Graduate School of Education has demonstrated this in his study of the schools that were part of the so-called Texas Miracle, which became the blueprint for the national No Child Left Behind Act passed in 2001. No Child Left Behind required states to create standardized tests for grades 3 through 8 and one year of high school. The tests vary from state to state and are used to determine whether schools are making annual progress in student achievement. Schools and school districts, in turn, use the test scores to evaluate teachers.

Deming compared the long-term outcomes of students who had been in Texas classrooms evaluated by the test-score metric with those not evaluated by it (during its pilot phase) and found that some schools likely held back or weeded out struggling students to jack up their overall performance on test scores. As a result, their students on the whole fared worse than their "untested" peers in college completion rates and earnings. This tended to happen in schools that were already above average and striving for a rating of excellence—they didn't want to lose out on getting a higher distinction. Meanwhile, the testing seemed to work out well for students in failing schools. Both kinds of schools, like hedge fund investors, appeared to act to avoid loss—though with very different results for students' lives. Other reports have documented how "teaching to the test" curtails student curiosity, and how it has even driven some teachers and principals to cheat by correcting student answers. The metric might work for organizations at the bottom of the heap, but not for those near the top. It took a long-term study to figure that out, but the test-score metric went mainstream long before.

Nowhere is there greater steering power given to myopic targets than in corporate America.

Today, a majority of executives who work at publicly traded companies admit that they routinely sacrifice future goals in order to meet the quarterly targets they set for earnings. This happens despite the fact that those targets are, in theory, supposed to be measures of their companies' long-term viability.

The corporate practice of projecting next quarter's profits for shareholders and Wall Street analysts (known as "issuing guidance") took off in the mid-1990s, when the U.S. Congress passed the Private Securities Litigation Reform Act. The legislation protects companies from liability when making predictions about how much they will earn in the next quarter. The law gave corporate executives the freedom to project high earnings to boost their near-term stock prices, which in turn boosts their bonuses and reflects well on their leadership in any given snapshot of time.

What happens, however, is that CEOs often manage their companies to ensure that they meet those earnings targets, lest they fall short of investors' expectations. By their own account, those leaders forsake key investments in long-term growth and invention in the process. The pressure comes not just from investors, but from leaders' own desire to meet the targets. Studies show that the more equity a corporate CEO has vesting in a given quarter, the more the CEO will slash investment to boost stock prices. The singular quest of the quarter becomes to fulfill what the CEO predicted.

Blair Effron, a banker and founder of Centerview Partners in New

York, which has advised General Electric, PepsiCo, Time Warner Cable, and many other iconic American corporations, told me that he routinely sees corporate executives making decisions driven by the quarter—cutting costs not solely for efficiency, and knowing it is at the expense of their own companies' future.

Surveys show that more than 80 percent of CEOs and CFOs are willing to sacrifice research-and-development spending—and half would delay starting new projects that would create greater long-term value for their companies—just to meet a quarterly earnings target. McKinsey & Company reported in 2016 that 87 percent of corporate board members and C-suite executives feel the greatest pressure to show strong short-term results, despite the fact that almost all leaders think a longer-time horizon would be better for both financial returns and for innovation.

Anyone who has faced a stream of impending deadlines can see why this happens—it's easy to sacrifice eating and sleeping well and to neglect important but not timely projects when constantly trying to meet immediate demands. By issuing quarterly guidance, corporate leaders set themselves up for a never-ending cycle of urgent deadlines. Their own metrics cue them to be perpetually in panic mode, much like the NASA engineers who erred in programming the Mars Orbiter. Or, in Cyril Connolly's terms, they become the rock pools that stay dry meeting the demands of the insatiable sea. Prediction, in this case, actually sabotages foresight.

Hitting the quarter target depletes the resources that companies need to do what is good for them over time. On a routine basis, companies buy back their own stock to boost its price and pay out dividends to shareholders rather than putting that revenue back into growing their companies. In 2015, the publicly traded companies in the S&P 500 spent 99 percent of their earnings on such tactics, leaving very little to invest in the future. There are a few notable exceptions, primarily technology companies that enjoy quasi-monopoly status and are pouring money into research and development focused especially in the areas of data analytics and artificial intelligence. These companies tend to be led by their founders, or have founders who retain significant voting power and ownership stakes.

In the headlines, we read about catastrophic cases of corporate fraud—the Enrons, Volkswagens, and WorldComs. Yet we are harmed in less sensational but more serious ways by the routine practice of executives sacrificing the future for the sake of the quarter. As a result of this behavior, employment becomes more volatile, invention stagnates, and companies fail to be supportive constituents for policies to address long-term problems such as climate change or K–12 education—concerns that don't pan out in the next quarter but that will affect the business down the line. Corporations also sacrifice the potential for people to earn greater returns on their pension funds or 401(k)s. When I think about this, it conjures the image of executives smashing piggy banks in the street and letting the change roll into the sewer.

A small but vocal group of people have begun raising the question of how to escape the magnetism of the myopic target in corporate America—specifically, how to dethrone quarterly earnings from its decades-long reign.

In the fall of 2016, several icons of the investment world gathered in the ballroom of the Hotel Sofitel in New York City to launch an organization they called Focusing Capital on the Long Term. At the helm was Sarah Keohane Williamson, a former investment manager who spent more than twenty years at Wellington Management Company and whose concern about corporate short-term thinking is motivated not by do-gooder altruism, but by money—and how much of it is lost for shareholders and for the economy as a result of narrow focus on quarterly earnings.

"On one end of the spectrum, we have savers like you and me, people investing in our pensions and retirement funds for the future, with access to capital," she told me. "On the other end of the spectrum, we have people with ideas who want to start businesses and grow them, who need capital. Both of those ideas and ends are very long term, but capital markets are short term. People are making bad decisions."

That year, the McKinsey Global Institute and Williamson's team began to look back at the performance of 615 publicly traded corporations in the United States over a span of fourteen years. The companies, none in the financial sector, were midsize to large, which they defined as those with market capitalizations exceeding $5 billion in at least one year of the time span. They deemed 164 of those companies to be long-term-oriented, measured by how narrowly they met or missed quarterly earnings targets and how their true earnings growth compared with what they reported on a quarterly basis. The data gave researchers an empirical way to tell whether companies diverted resources from future needs to meet near-term targets. Among other net gains, they found that the companies that had overcome focus on quarterly metrics earned 47 percent more revenue than others in their sectors from 2001 to 2015.

Most important, perhaps, is that they found widespread benefits of being oriented toward future growth. The long-term firms spent on average 50 percent more on research and development than the other firms and added almost twelve thousand more jobs. The McKinsey research team estimated that, had all U.S. companies acted like the long-term companies in its study, the country would have added another $1 trillion to its gross domestic product during that period. Society has paid a high price for the corporate distraction by metrics.

It's not only big companies that suffer from myopia. Silicon Valley entrepreneur Eric Ries has called out the perils of what he dubs "vanity metrics," which start-up companies use to gauge their success at a given moment. He points to technology companies that rely on page views, downloads, and numbers of users of website services as measures of progress. Start-ups, in his view, have a tendency to credit spikes in these numbers to wise business decisions—when they might just be seasonal or weekly fluctuations in business, similar to what Disneyland sees in attendance numbers on weekends and school vacations. By contrast, dips in these numbers can be taken too seriously—just as a hedge fund investor might overreact when she looks too often at her portfolio. Organizations' own metrics, not just those of investors, often obscure foresight.

Ries says the most successful, enduring companies choose metrics that mean more over time—instead of gross numbers of page views, for example, they count the number of times a customer returns to a site per day, which tells them how much she likes the product and how loyal she is, better indicators of future reach. Ries points to Facebook as an example of a company that used the latter kind of metric, even in its early days. The strategy of earning loyalty has made it hard for people to quit using the social media platform, even amid concerns about data privacy, false news propagation, and the company's ethics.

The specific metric that makes sense for an organization to use depends on what its objectives are. Repeat purchases and market share may make more sense for retail companies, while frequent visits or use might be better gauges for apps or businesses that rely on advertising. For news organizations, the time readers spend on a story or number of times they return might be more meaningful than the number of clicks, even if the latter wins advertisers. But no metric is a substitute for honing judgment.

Ravenel Curry has an air of gentleness and humility, and a lingering drawl that summons his youth growing up on a South Carolina peach farm, despite the intervening decades spent in New York and New Jersey. He does not take credit for knowing better than others who stayed in the market during the first dot-com bubble. He says it is simply the only way he likes to invest.

Like all investors, Curry and his team at Eagle aim to buy a stock at a bargain and sell it at a higher price. And like all investors, they face uncertainty about what the future will bring. That uncertainty grows as they look further out in time. No one has a crystal ball.

To outperform the market, investors need to defy consensus. Eagle's strategy is to look for short-term headwinds keeping a stock price low that are likely to disappear or become irrelevant over the course of years—as opposed to hours, days, or weeks.

Investments that look bad here and now are, as a result, often the ones that pique Eagle's interest. The firm's strategy is to buy stock in publicly traded companies when it is undervalued by most investors, perhaps because earnings are low, costs are high, or there's a perceived short-term issue, like a fear of rising interest rates or speculation about a peak in commodity prices.

Some investors call the Eagle approach "time arbitrage," but it might also be seen as profiting from others' shortsightedness. The firm seizes upon the fact that Wall Street traders and their sell-side analysts often overreact to irrelevant metrics and fail to place adequate value on stocks' potential future growth. A company like Amazon, for example, turned off many would-be shareholders and traders for years, because its costs were skyrocketing and profit margins shrinking. Meanwhile, the company was building an empire by reinvesting its earnings to develop new products and grow its market share as a retail juggernaut.

Eagle Capital is not the only investment firm that has amassed a fortune by looking past myopic metrics. The billionaire investor Warren Buffett famously ignores short-term noise for the truly long term— sometimes preoccupying himself more with the eventual risk of nuclear annihilation than with the next few years. In 2008, Buffett made a public $1 million bet against the leaders of a hedge fund, Protégé Partners. He predicted that in the next ten years, the S&P 500, and the low-cost, passive index funds pegged to it, would outperform a selection of leading hedge funds chosen by Protégé that charge their clients high fees. Protégé bet on the hedge funds doing better in that decade, which would have proven the value of their line of work. Buffett won the bet, and donated the proceeds to Girls Inc., a charity in Omaha. He proved that tuning out noise entirely could be more profitable than tuning it in.

Like Buffett, Seth Klarman, the financial mastermind behind Baupost Group in Boston, has had extraordinary success betting on companies he thinks will gain long-term value, not simply trade at a higher price in a week. Baupost manages around $30 billion, and a copy of Klarman's out-of-print book on investing against the grain of Wall Street can fetch more than $1,000 on Amazon.

On the eve of the Great Recession of 2008, Eagle Capital did not own stock in any of the big banks. The banks were overvalued, the Currys and their investment team believed, because most traders were looking at the strength of short-term returns. The banks seemed to be taking a lot of risks to make high quarterly earnings. In private, people working in the risk departments of big banks complained of not being heeded. If you looked, as Eagle did, beyond the quarter or year, the odds of a blowup grew.

We all know how the story ends. The stocks of major banks crashed as expected. What happened after the crisis captured the Currys' interest: The banks' share prices continued to be depressed well into 2009. That's when Eagle began to investigate. In 2011, after careful deliberation, the firm started buying large stakes in a few banks—Goldman Sachs, Morgan Stanley, and Bank of America.

The Eagle hypothesis was that it might be a good time to buy stock in certain banks because the market was still reeling from past events and not adjusting to the opportunities of the future. Bank executives who had taken the helm after the financial crisis should be averse to taking big risks, because they knew the markets and regulators were watching them with greater scrutiny. Incoming CEOs and CFOs of banks after the financial crisis had the motive to root out problems or high-risk products in their companies right away—while they could still blame any wrongdoing or bad ideas on the previous executives.

This was the view of Boykin Curry, the son of Ravenel and Beth, who joined the family firm in 2001 after stints in management consulting and hedge fund investing. (Ravenel is still at the firm; Beth passed away in 2015.)

To make such bets on the future amid uncertainty, the team at Eagle uses what Boykin calls reverse stress tests.

The test works like this: When the team is thinking about investing in a particular stock, Boykin Curry will propose a hypothetical scenario. "Imagine it's five or ten years from now, and we are looking back and feel we made a grave mistake buying this stock or not buying it. What are the things that might have happened to make us feel that way? How likely are each of those things to play out?"

The reverse stress test helps the investors consider a wider range of risks and opportunities in the future. The practice is similar to a strategy used by Google X, the arm of the tech giant that invests in "moonshots" to solve big societal problems that affect millions of people. Google X was responsible for the company's ambitious self-driving car project, and a newer effort called Project Loon to build an internet powered by balloons in the stratosphere to connect people in remote and rural areas of the world.

By design, the projects Google X pursues are unpredictable and high stakes; they are aimed at solving daunting problems and require millions of dollars of investment over time. It's important to know as soon as possible whether an investment will succeed or fail. Google X's director, Astro Teller, gleefully boasts that every day his team tries to kill off projects. "We don't want to wait till the failure happens to learn from it."

Before ever launching a project, Teller's team will run what he calls "pre-mortems"—as opposed to postmortems. The idea of a pre-mortem is to predict why a given idea or project will fail. Each person on the team is encouraged to make such predictions by writing down possible risks or problems. The team votes up or down potential threats, to gauge them against likely reality. If enough people are persuaded by a threat, the team will kill off the project before even starting it. Team members get rewarded, rather than ridiculed, for coming up with these reasons to terminate ideas. The rewards are often as simple and cheap as high fives or hugs.

In the 1980s, Deborah Mitchell, a professor at the Wharton Business School at the University of Pennsylvania, studied the technique of explaining a future event as if it had already happened—which she dubbed "prospective hindsight." For example, you might imagine that you have already hosted a successful party, and then state all the possible reasons—a thoughtful menu, a group of lively guests, good music—that it "turned out" so well. In contrast to the more common practice of describing *what* will happen in the future, prospective hindsight requires assuming something already happened and trying to explain *why*. This shifts people's focus away from mere prediction of future events and toward evaluating the consequences of their current choices.

Mitchell thought this practice could help groups make better decisions by opening their eyes to unexpected ways things can go wrong or right, and alerting them to uncertainties. She believed prospective hindsight might help an organization uncover errors, such as those that led to the 1986 Chernobyl nuclear disaster in the former Soviet Union. The strategy of fast-forwarding time and explaining how we got there, echoed today by the pre-mortem, can aid our foresight.

The majority of money managers on Wall Street interact with company executives via group calls, where they—or their sell-side analysts, who are not shareholders and benefit from more trading—try to get information about the next quarter's numbers. They put these numbers into a model that then guides pricing and trading decisions. Eagle Capital's approach to making an investment decision is far more time-consuming. Before investing in Morgan Stanley, for example, the team spent days talking to the management team and looking at their books. They talked to the head of risk management at the bank, the head of trading, and the head of investment banking. They wanted to know how the bank's leadership would avoid risks and gain greater market share over time.

Eagle investors admit to being pains in the ass when it comes to buying a position in a company. But since they hold on to investments for several years, they argue it is worth an executive's time answering up-front questions. The nature of the questions they ask is also different from the nature of questions many other investors ask. They want to get a sense of how executives and managers think and what their values are.

They pose hypotheticals: What would executives be willing to do or not do to make their quarterly earnings projections? How much do they care about innovation or about growing the company's share of the market? How far into the future do they gaze when making decisions?

Sarah Keohane Williamson, the former Wellington manager, tells me that most investment fund managers are afraid of underperforming for a few quarters lest they lose clients. Investors and their clients try to avoid immediate loss.

Investment managers like the Currys who hold their stocks for several years have to stay out of bubble markets when prices are rising and keep their nerve when prices of their stocks dip in the short term or earnings weaken. Over the course of years, there will likely be many periods when it will look and feel like they are losing money, or when they are tempted to join a manic market. Investment firms and other organizations need techniques to avoid the twin perils of indulgence in instant gratification and acting rashly to avoid immediate loss. They need to exercise collective foresight amid the noise.

How does anyone distinguish between signal and noise—between a veritable canary in a coal mine or harbinger of progress, and a mere distraction?

Boykin Curry told me that when Eagle investment team members see the share price of one of their investments dive, they kick in a prepared response. They start by reminding themselves of the original reason for buying a stake in the company. For example, they bought stock in Microsoft in 2006 because they thought the company could become a leader in cloud computing. Three years later, the stock price had fallen by more than 40 percent. The Eagle team investigated. They talked to the management at the company and experts outside the company to find out the reason. It turned out the price had dropped with a decline in personal computer shipments and prepackaged software sales. "That had nothing to do with why we bought the stock," Boykin said. The team's plan was to ignore the noise in the market about Microsoft, and they even bought more stock.

"On the other hand, if we had found out Microsoft was selling off their cloud business, we would have lost our reason for investing in the company. But hopefully, that business decision would boost near-term earnings once the company made the sale, and we could sell at a higher price than we bought at," he said. They avoided distractions of the moment by returning like a broken record to their original rationale for believing in the stock. You might call this a North Star tactic, calling on people in an organization to habitually look up from daily minutiae to reorient themselves to their ultimate destination.

In 2016, I attended the Stanford Directors' College, a sort of five-day summer camp for people who sit on boards of publicly traded corporations. At lunch one day, I sat next to Roger Dunbar, chair of the Silicon Valley Bank, who stood out among the corporate suits because he looks more like a roadie for a rock band than an executive. Dunbar seemed like a Valley savant, cut from the cloth of the 1960s. He has advised and helped launch many technology companies and formerly served as the global vice chairman of Ernst & Young. He told me that when he hears company executives or board members responding to short-term noise with outsize reactions, he likes to pretend he is lost. He'll ask CEOs at board meetings, "What was our long-term strategy again?" as if he has forgotten it. Sometimes, he says, company leaders suffer from being too smart—analyzing every piece of data that comes their way—instead of asking simple but pivotal questions. It can help, in his view, to have a board member who is not afraid of sounding naive or even a touch senile. I see this as another way of reminding people to look up. It's a role that boards, dedicated teams, and advisers can play.

In the financial world, most investment managers are rewarded on the basis of a fund's annual performance, and given bonuses based on how their stocks fared and what they earned for clients in that year. An Ernst &

Young survey showed that an average of 74 percent of fund managers' compensation comes in annual cash payments. This means incentives that would shift their gaze to the future, such as equity shares, deferred cash, and stock options, make up less than a quarter of their pay.

Williamson says such pay structures often get in the way of investors looking at future risk. Like the flashing lights and melodies of a slot machine, they reward the pursuit of instant gratification. They also motivate people to act on fear of immediate loss. No one wants to miss a bonus. What's troubling is that annual bonuses are fundamentally at odds with the reason most households put money into investment funds—for the long-run goal of financing their future retirement. To have that money managed by people with yearly targets betrays the interests of the people who stand to gain or lose from their decisions.

Eagle Capital does not give its portfolio managers annual bonuses. Nor do its managers get annual performance reviews. Their compensation and equity in the firm grow as the firm grows, over time. Boykin Curry says this is because he does not want the managers making decisions about investments "just because it's the end of the year and they want to buy a Ferrari." A year's time, in the Eagle philosophy, is also simply too soon to see how an investment decision has played out. The firm is both cooling the heat of temptation and smoothing out the prospect of loss by taking away these near-term rewards—as if silencing the slot machines and removing the free perks from a casino. A portfolio manager is thus trained to see that this year's performance will not make or break him.

In essence, the Eagle organizational culture is meant to encourage foresight. Staying small makes this possible. The entire investment team at any given time is six or seven people, which gives the Currys the luxury to look beyond how the team is faring in a given instant. They can see how each person thinks and whether their reasoning in making investment decisions eventually plays out. The firm invests in only about twenty-five to thirty-five companies at any given time, making it feasible to keep close tabs on the decisions being made, the leadership, and the industry trends for each of their stocks.

It's worth noting that the Eagle Capital team does not eschew metrics altogether. It's just that they look past the typical investor's metrics, instead focusing on indicators that they see as proxies for future opportunity—and those that signal risk. Myopic metrics are replaced by milestones, which reflect whether or not the theory of a company's growth is playing out.

As an example, when other investors were trying to figure out when interest rates would rise and therefore affect the stock value of big banks, Eagle was looking at the closing values of derivative contracts that Morgan Stanley had put in its books ten years prior, a lagging indicator of whether the bank's team was accurately reporting the value of its assets or still taking undue risks. Instead of choosing measures that reflect only the immediate prospect of success, milestones like these test assumptions, so that one can adjust a view of the future along the way.

I asked American microfinance expert Daniel Rozas, who foresaw the 2010 microcredit crisis in India, if there were any metrics that can serve as better canaries in a coal mine in his field than loan repayment rates. He developed a system to gauge whether there is a credit bubble growing in a given country. It involves comparing the amount of potential borrowers with the number of loans to determine whether too many loans have been dumped on borrowers in the country, and watching the rate of loan growth. He also looks at government regulations and whether they curb excesses among companies, and the amount of transparency in the market. He does field surveys and interviews with borrowers to see if the metrics are aligning with the reality on the ground. A single metric, in his view, always runs the risk of deluding us.

Dan Honig, a political scientist at the Johns Hopkins School of Advanced International Studies who researches foreign aid, believes that metrics can be useful when an organization has a simple, concrete goal such as building a road. The trick is for the measure to be tightly linked to what the organization actually wants to accomplish. "I used to be an apartment maintenance worker, and if I patched a hole in a wall, it didn't matter if my boss liked me, because at the end of the day I patched a wall," Honig told me. "It was verifiable. Building a road or delivering a vaccine works that way—it's something you can observe. Counting whether someone paved five miles of road per month or put a shot into a hundred people's arms is a close proxy for what the organization wants to achieve." For the more complex undertakings of organizations, however, numeric targets are often far removed from actual goals and more likely to deceive. In those instances, Honig says, managers are better off using their judgment to evaluate progress. It's a common mistake for organizations to attach simple metrics to nuanced goals such as educating children, reforming the justice system, or growing an innovative business.

Some CEOs today are choosing new metrics that they think better align with long-run goals. When Paul Polman took the helm of Unilever, the European megacompany behind brands including Dove soap and Lipton tea, he aspired to see the business endure for a century beyond his tenure—and to carefully manage supply chains for natural resources it uses such as palm oil. After Polman became CEO in 2009, the company stopped predicting quarterly earnings because of his belief that hitting targets cannibalized efforts to think about long-range concerns, including whether the company was destroying the planet. Other companies, such as Coca-Cola and Ford, also stopped projecting by the quarter to signal that they wanted patient investors.

Unilever has tied Polman and other top executives' annual compensation to metrics that align with long-range objectives such as reducing the company's carbon emissions. In 2016, Polman received a $722,230 bonus for progress on the company's sustainability goals. The Dutch company Royal DSM adopted a more ambitious reform, tying half of short-term bonuses to sustainability goals for more than four hundred of its employees.

Jeff Bezos, the founder of Amazon, defined his own metrics. For nearly twenty years, Amazon turned no profit for shareholders while building its retail and cloud computing empire. At the height of the dot-com bubble, in 1997, when internet companies were trading on froth, Bezos penned a letter to the shareholders of Amazon in which he explained that he had a long-term vision for the company. In this letter, often cited today by investors for its audacity, Bezos explained how the company would make decisions to grow over time and become a market leader in the future. "We believe that a fundamental measure of success will be the shareholder value we create over the *long term*," he wrote. He never said exactly how long that would be.

Bezos spelled out alternative metrics in the 1997 letter to replace the quarterly profit and the near-term share price: Amazon would measure customer growth, to show how much of the market the company was capturing. It would also gauge customer loyalty by repeat purchases; and it would measure the strength of the brand. He articulated his own version of a North Star strategy and how it should be judged—by tracking the company's progress on its own metrics, not Wall Street's.

Bezos also explained up front his philosophy of making decisions for the future by describing how he would deal with specific trade-offs. The company would choose growth over current accounting measures, and pay its employees more in stock options than cash to make them invested in the company's future. It would aim to keep cash flow high and reinvest that cash in future growth.

Amazon reported low to zero profits for nearly two decades after that letter, but not because the company did not bring in money. Rather, Bezos chose to reinvest the billions of dollars of revenue that came into the company in new businesses and technologies instead of just recording this cash as high earnings to satisfy Wall Street. The strategy worked spectacularly well, as Amazon is now one of the most valuable companies on the planet. Not every new venture has taken off, but a few large wins have made the strategy financially worth the costs. Recently, one of those investments, the company's cloud computing platform Amazon

Web Services, has driven up revenue and become an indispensable tool for companies that collect and analyze large stores of data.

Amazon's marketplace dominance concerns many people who believe that more competition, and smaller businesses, would be better for consumers and for the economy as a whole. Others laud the way the company has changed the experience of shopping. I agree with some criticisms of the company's practices, and am conflicted about my own addiction to a service that uses so much packaging. But I don't believe such concerns obviate the opportunity to learn from Amazon's example about foresight—lessons that can be applied by other organizations.

The burst of the late-'90s dot-com bubble woke up many investors to the reality that some technology companies would never actually turn profits, and were trading at high prices based on pure speculation. After the crash, Amazon had to distinguish itself from those technology companies and convince investors that there was something real to come. The astronomical sales growth—an alternative metric to quarterly earnings—helped Bezos make that case to investors.

Most corporate CEOs today pad their earnings rather than make down payments on future business. Bezos does the opposite. At various times, Amazon's share price has been depressed as a result of business decisions that led to disappointment about a particular quarter's numbers. Today, many early investments have begun to show their high profit potential. In September 2018, Amazon became the second publicly traded corporation in history, after Apple, to be valued at $1 trillion. The irony, of course, is that this is a company that encourages its customers to make ever-more-impulsive buying decisions. It seems that long-range thinking, despite being part of the Bezos gospel, is not necessarily his moral agenda, but rather an ethic he applies to his own interests.

Paul Polman and Jeff Bezos remain exceptions in Western corporate leadership: Most company executives don't believe they can pull off such strategies, or don't want to take the risk of trying and failing.

Investors, for their part, could also choose alternative metrics to quarterly profits to gauge the long-term prospects of companies. They can also change the metrics by which they judge their own success. Berkshire Hathaway uses the rolling five-year performance of the Standard & Poor's index as its benchmark, instead of year-by-year performance versus the S&P, to gain a longer-range perspective. But most investors have yet to adopt such practices.

John Henry, the American folk hero, drove a steel drill into rock with brute force. In the popular telling, Henry goes head-to-head in a competition with a steam-powered hammer that is poised to replace him. He rises to the challenge, hammering his drill to set explosives to build a railroad tunnel. He outmatches the steam hammer, which breaks, while his will and his back endure.

It often goes overlooked, however, that in the legend John Henry dies from exhaustion after this feat of strength. The race with the machine claims his life. Yet somehow the story has still been construed as a triumphant fable of human prowess.

There's a flaw in the moral message that the human hand and spirit can always prevail over the machine. Humans are likely, in fact, to be out of work if we cling to doing jobs that machines can do more efficiently or with less risk. An outboard engine beats an oar. A computer is faster than longhand calculation. An electric lawn mower is much more efficient than a scythe. We're better off doing things like rowing a boat or hand-writing a letter for the pure joy of it, not as occupations.

As we collect more and more data, and as computers become more sophisticated in learning from those data to make decisions, machines are competing with and surpassing our capabilities in many arenas. Machines of prior eras replaced only the most physical of human labor. Leaps in artificial intelligence today have made the machine a rival for tasks requiring human cognition. Autonomous vehicles are poised to displace truck drivers. Computers can already outperform lawyers in amassing documents relevant to a given case, and radiologists in detecting tumors from MRIs.

Part of Eagle Capital's edge comes from doing what machines can't do well—at least not yet. Since the 1990s, many investors have been heavily guided by computer models. Sophisticated analytics help make sense of vast stores of data on past transactions, current and past stock prices, commodity prices, and forthcoming earnings to reveal opportunities to buy or sell. In the view of the Currys, this approach to investing has become more and more competitive, and crowded with firms that have access to the same data and marginally better or worse machine learning and data analytics tools.

"We have to exploit the patterns only humans can see," says Boykin Curry. "We have no competitive advantage when everyone has the algorithm." I see Eagle's approach as adaptive in a way to how technology is displacing human talent. At least for the next few years, it's hard to imagine a device sharing the investors' penchant for asking probing, annoying questions or their habit of looking beyond the dataset with reverse stress tests.

A few years ago, I met Demis Hassabis, one of the world's leading artificial intelligence researchers, at a rodeo club in Montana. Hassabis is a neuroscientist who studied human memory as a doctoral student in England and then founded a company called DeepMind, now a subsidiary of Google. Hassabis heads up the company's ambitious efforts to make machines learn better than humans do. The company made headlines in 2016 when its AI system AlphaGo beat the best human player alive at the thousand-year-old Chinese board game Go. During the game, the machine improvised, using the accumulated data of many games of Go to figure out new moves to win the game. The feat showed that a computer can learn from situations and make intuitive and creative decisions, not just follow the rules given to it by humans.

Hassabis points to this victory as evidence that artificial intelligence is outpacing researchers' expectations. But he believes it will be a long time before machines gain the ability to make value judgments about the future.

Artificial intelligence can scan for and recognize patterns in data, arguably better than humans can. Computers can learn from large datasets that people cannot digest in one sitting—or even one lifetime. Humans tend to learn from their own experiences and those of others. Hassabis sees robots taking over roles in our society where there is an edge from extracting knowledge from large stores of information and making sense of it. In hospitals, for example, machines might one day soon make better and faster diagnoses of patients' illnesses than doctors, because they can learn from diagnoses made by millions of doctors about millions of patients. Self-driving buses will one day be guided by computers that can learn from hundreds of millions of driving dilemmas. In investing, machine intelligence is already showing its edge over humans in parsing the copious information that guides short-term trades. Goldman Sachs has already automated its cash equities trading, replacing all but two of six hundred human traders with a cadre of computers supported by engineers.

People still have an advantage, however, when it comes to understanding underlying factors that might not adhere to past precedent or trends, and in detecting deep shifts. Howard Marks, a widely respected value investor who heads Oaktree Capital Management, has pointed out that quantitative investment funds' reliance on computer models can be their Achilles' heel. The models eke out profits by relying on patterns that held true in the past. "They can't predict changes in those patterns; they can't anticipate aberrant periods; and thus they generally overestimate the reliability of past norms," he writes.

A computer model using solely historic and real-time data might underestimate the abilities of a company's new CEO to boost the share value over time if she lacks the pedigree of past successful executives, for example. That same CEO, however, might be perfectly suited to the unwieldy times ahead. A human has the ability—at least hypothetically—to ask deeper questions about trends that are outside the typical dataset and to look beyond the numbers.

By contrast, relying on a computer to make decisions will tend to overemphasize metrics, because most programs rely primarily on numeric inputs. And although a computer can perform tasks far more quickly than humans, such as buying or selling stock by the millisecond, such high-frequency decision making can lead to catastrophe when an algorithm fails to account for a risk that a human would detect. A series of automated decisions can snowball quickly to cause a firm or even the entire stock market to crash.

In a world where AI plays more prominently, Hassabis believes humans will still outperform machines in complex and empathic roles. The human edge will also arise from our ability to look beyond data points, and beyond simply hitting targets that we can program machines to meet—like winning a game.

We have not yet adapted to this emerging trend. Right now, in medicine, the emphasis for doctors is on making the correct diagnosis—hitting that target in front of them. Less emphasis and fewer resources go to the long-term work of caring for ailing patients, and helping patients navigate the pros and cons of treatment options. Similarly, in investment, the emphasis is on making a marginally better prediction of next quarter's numbers and stock price, rather than gauging the long-run value of a business. We have yet to recognize that foresight trumps information in the game we are playing with machines.

It would be a bitter irony to remain entranced by myopic metrics, gearing ourselves up to hit immediate targets at a time when technology and our economy are evolving to privilege humans for being visionary, empathetic, nuanced, and strategic. In the future, the human edge is going to come from what we value and from our judgment, not from going head-to-head with machines to parse facts.

The success of firms like Eagle, Baupost, and Berkshire Hathaway shows that it pays off to be patient when everyone else is locked on immediate results.

But sometimes, organizations find they have to follow the herd of the myopic in order to survive.

Fishing businesses have faced this reality for centuries. Even for fishermen who aspire to pass on their trade to their children and grandchildren, it is difficult to avoid harvesting fish until their populations are wiped out. Fishermen race to get what they can before others get theirs—even if that means everyone loses in the long run.

Until recently, Buddy Guindon was one of those fishermen. His commercial fishing business, based in Galveston, Texas, pursued the reef-dwelling Gulf Coast red snapper. For more than fifty years, fishermen overharvested red snapper to the brink of extermination. By the early twenty-first century, the fish population had declined to just 4 percent of its historic levels, and was on the verge of disappearing forever.

The U.S. government historically tried to protect the Gulf Coast red snapper, and other fisheries in federal waters, by setting an annual catch limit and allowing fishing on certain days each season until a yearly quota was reached. In a single trip, the Gulf of Mexico fishermen were allowed to fill their boats with up to two thousand pounds of red snapper—a daily target. During these designated "derby days" a few times a year, commercial fishermen raced one another to get as much fish as they could. They tried to beat others to an ever-dwindling supply.

Guindon says he, and other fishermen he knew, fished like pirates under this scheme. They left weddings and their kids' baseball games to get to the fish first on derby days. They even stole out to sea under the cloak of night. To stay under the daily cap of two thousand pounds, they threw overboard smaller fish they caught in favor of larger fish for the market. Yet the fish brought up quickly from the ocean's depths typically did not survive when returned to the sea. Miles of dead bycatch trailed the boats.

Derby days concentrated all the fishing at the same times of year. During these days, fishermen killed vast amounts of fish, including those they brought ashore legally and illegally, and those they dumped at sea as dead discards. This harmed red snapper reproduction and kept the populations from rebuilding each year. As the fishery declined, commercial fishing boats would go farther and farther offshore for longer periods of time in search of fish, even in storms, to make sure they could get as much of the harvest on the open fishing days as possible. They wasted fuel, time, and money. They fished recklessly.

On the dock, the market for red snapper suffered, too. The fish was available in a glut on derby days, fetching low prices, and not available at all during other times of year. Chefs stripped the species from their menus, and fishermen reaped fewer and fewer profits while putting in more effort to fish at greater risk to themselves and their crews.

Even if fishermen like Buddy Guindon wanted to fish moderately—with more of an eye to the future good of the fishery and their own families—it was not in their interest to do so under the derby days policy. Everyone else would get to the fish first and leave nothing for the fishermen who were cautious. "It motivated me to destroy my own fishery," Guindon told me. "You can call me a crook if you want to, but that is what the regulations did."

Fisheries have long been seen as classic exemplars of the "tragedy of the commons." A seminal essay by Garrett Hardin from 1968 describes this as the conundrum of taking care of shared resources. With each person pursuing her own self-interest to harvest fish or dig for gold, a race ensues to use as much of the resource as possible before someone else gets to it and destroys it. No single person will conserve the resource, because they all fear that the neglect of others makes doing so fruitless. Nobel Prize–winning economist Elinor Ostrom pointed out that Hardin was actually quite late to the party of thinkers who came up with this idea. Aristotle wrote in his treatise on politics, "What is common to the greatest number has the least care bestowed upon it. Everyone thinks chiefly of his own, hardly at all of the common interest."

Viewed another way, however, fisheries including the Gulf Coast red snapper have suffered more from what I call a tragedy of time horizons. Over time, overfishing, regardless of the short-term profit, is not in the self-interest of most fishermen. Their survival, livelihood, and way of life are tied up in the health of the fishery. Fishing is not simply an economic activity for the majority of fishermen, but a cultural identity that defines them and their communities. It is also a legacy that many fishermen aspire to leave for the next generation. This is something I have observed consistently in visits to and studies of fisheries around the world over the past decade.

Buddy Guindon and Gulf fishermen like him saw their fishery's destruction in progress, and saw the damage they were doing to their own future livelihood, as well as their community's future. But they did not act on their concern for the fate of the fishery, because the policy of derby days reinforced their recklessness.

I believe many situations we view as tragedies of the commons might also be viewed from the lens of time horizons. When the ocean or air gets polluted by individuals, either they or their children eventually suffer. In making these decisions, people are biased toward taking an immediate payoff despite the great long-run cost. If more people placed greater weight on their future self-interest—where it also happens to be the collective interest—we could overcome some tragedies of the commons. Ostrom has similarly argued that there are ways we can do this—that the tragedy is not inevitable.

In 2007, the Gulf Coast red snapper fishermen found a salve for their wounds. A new approach for managing the fishery—called a catch share—had been brought to their attention by a conservation group, the Environmental Defense Fund. They voted to adopt it. The Gulf of Mexico's regional fishery management council, established by national law in 1976, approved the scheme as a way to rebuild the fishery.

Under a catch share system, fishing businesses each get a guaranteed portion of the annual quota of fish that can be harvested. Shares in the Gulf Coast were allocated based on history—which businesses had landed red snapper for the previous fourteen years. The fishermen can now choose to fish at any time of the year to fulfill their share, or they can lease their stakes to another fisherman, who will pay them for the right to fish that year. They become shareholders in the fishery, but they don't buy and sell their shares rapidly like Wall Street traders do. There are higher barriers to trading shares; it takes skill, a boat, and expensive equipment including hydraulic reels to be in the red snapper business. Most of the fishermen who have shares keep them for the long haul, acting more like founders and owners than like stock traders. Unlike Western ranchers given water rights that they will lose if they do not use enough water in a given year, the fishermen are ensured their shares over time, so there is no motivation to overfish today.

The catch share scheme for Gulf Coast red snapper has had extraordinary success. It brought the fish population back from the brink of collapse, restoring stocks nearly decimated fifteen years ago.

The spawning population of red snapper has tripled since 2007. As it rebounds, the bounty of each commercial fisherman grows. The amount of red snapper commercial fishermen are permitted to catch each year by law has more than doubled, and as a result, the revenues earned by fishermen including Buddy Guindon have doubled.

The fishing season is now spread out over the course of the year, which puts less pressure on the fish stock at any given time. It also relieves fishermen of the need to risk their lives at sea in inclement weather. Guindon reaps more profit today because his costs have fallen. He can bring aboard 10,000 pounds of fish in a trip that once limited him to 2,000 pounds because he is allowed to catch the fish anytime he wants over the course of the year. He discards 80 percent less fish because of this, and he can fetch higher prices at the dock because he is not bringing in his fish on the same days as everyone else.

Catch shares let Guindon and other fishermen be the long-term investors that they aspire to be instead of setting them up for a race to catch every fish they can right away.

A study published in the journal *Nature* in 2017 compared thirty-nine fisheries managed by catch shares with fisheries in the United States and Canada that had not adopted the approach. The researchers found strong evidence that catch shares stopped the race to fish in each of the regions where they had been adopted. Garrett Hardin might not have been able to imagine what has happened in the United States during the past two decades; the number of overexploited fisheries has declined to the lowest level since the 1940s. About two-thirds of the fish brought ashore today from federal waters are caught under catch share schemes, and more than forty U.S. fish stocks have rebounded since the year 2000.

The success of catch shares shows that agreements to organize businesses—and wise policy—can encourage collective foresight. Programs that align future interests with the present can, in the words of Buddy Guindon, turn pirates into stewards.

Many CEOs today feel like fishermen out to sea during derby days—they are in a perpetual short-term race to make their quarterly numbers look good and get high valuations that in turn boost their stock prices and earn high returns for shareholders, and keep their boards happy. They need an impetus to shift their companies' culture toward collective foresight. A lot of executives say they want to weigh the future more heavily in their decisions, but their investors won't let them.

In 2017, I met up with Ron Shaich, the founder and, until early 2018, the CEO of Panera Bread, one of the most successful restaurant companies over the past twenty years. It was a few months after he took his company private, exiting the capital markets. Shaich saw the company's competitive edge in making long bets on future trends in restaurants, such as the demand he anticipated for "fast-casual" spots like Panera that serve soups, salads, and sandwiches at low prices. His business mantra is to look ahead five years at where he wants the company to be, and to work backward. In his years leading Panera as a publicly traded company, he often found the pressures of Wall Street to be counter to these aims. He had run-ins with activist shareholders who wanted him to abandon investments in technology—bets that paid off eventually in rising to-go and delivery orders—because earnings had been flat for a few years. Other shareholders pressured him to buy back shares of the company to boost the stock price.

Shaich calls himself "long-term greedy, not short-term greedy," because he finds that he can build more value for a business by focusing on points far beyond the quarter. He likes to think ahead about risks, too, telling me "the time to worry about a heart attack is not on the way to the hospital." While at the helm of his publicly traded company, however, he routinely found himself spending 20 percent of his time reporting on the company's recent activity and its next quarter's activity, instead of focusing that time on where he wanted the company to be positioned in a few years. He also saw an increasing compression in the time horizons of his shareholders. When he first took Panera public in 1991, he says, more than half of its shareholders held on to their stock for a year or longer. Twenty-six years later, half of his shareholders held on to the stock for less than a month. Even though some long-term investors stuck around, the stock price was dictated by hedge funds making fast trades, interested in the performance relative to competitors' last quarter and not in the company he was building.

In 2017, Shaich courted patient private investors to buy Panera and take it off the stock market. The investors, led by a group called JAB Holding Company, paid $7.5 billion for the company. They identify themselves as investors interested in buying companies with future growth prospects. As it is currently structured, JAB, which in 2019 was revealed to have Nazi origins, professes to have a time horizon of centuries. It steers business decisions on the basis of multiyear time horizons, not quarterly ones.

Many company founders and leaders choose never to enter public markets for the express purpose of staying true to their values as a company. Jamie Dimon, the CEO of JPMorgan Chase, believes that staying out of public markets has become more attractive to companies that don't want to be subject to the myopia of Wall Street, and that's why initial public offerings are on the decline. Some companies, like Plum Organics, King Arthur Flour, and the apparel company Patagonia, even become what are known as benefit corporations, a designation that makes them legally accountable not just for financial returns to shareholders but for meeting specific environmental and social criteria. Benefit corporations tend to attract investors who are deeply invested in the brand and the company, and care about a business's broader impacts on society. Most are privately held companies, and their growth can be constrained by their access to capital.

Neither staying private nor going public is a guarantee, however, of whether a company will orient toward long-run growth or stay focused on short-run profit. For most companies, it is a question of leadership and the tools that leaders use.

In the twentieth century, corporations invested far more in early-stage, long-term research and development than they do today. Basic research investments at IBM, GE, Xerox PARC, RCA Labs, and AT&T's Bell Labs in that era brought forth the transistor, solar photovoltaics, and lasers and infrared imaging. Today, companies' R&D investment has shifted to late-stage technologies that deliver new products more quickly, and U.S. private-sector R&D as a share of economic output has fallen. The country that used to be the undisputed global leader in research investment is now ranked tenth in the world. The recent boost in overall R&D spending can mostly be attributed to just a few technology companies such as Amazon, Intel, Alphabet, Apple, and Microsoft, while such investment has dropped in other sectors.

Lynn Stout, the late Cornell Law School professor, attributed the decline in long-term investments at American corporations to the rising influence of shareholders with short time horizons, who behave less like the long-term owners or corporate boards of a previous era and more like speculators who want to extract as much cash as they can out of a company today. "Shareholder primacy" was abetted by rules put in place in the early 1990s by Congress and the Securities and Exchange Commission, she wrote. These reforms led to companies tying CEO pay to share price and quarterly earnings per share—and gave shareholders like activist hedge funds the ability to organize campaigns to make corporate boards respond to their immediate demands.

The mantra for corporate America coming into the twenty-first century has been to "maximize shareholder value." That idea does not necessarily have to encourage myopia, but it has come to mean maximizing value this minute rather than over time. After all, you could have shareholders who demand foresight and investments in research and development and new markets. The ease of trading shares, abetted by technology, and the impatience of investors, however, has meant that shareholder primacy coincides with shortsightedness. Shareholders do not stick around like true owners of stakes in companies, to ride out brief dips in stock prices. They respond to near-term targets, just like the New York City cabdrivers who focused on their daily earnings at the expense of their monthly goals. Uncertainty and fear play into it, too. In a 2015 article, "The Corporation as Time Machine," Stout wrote:

> Consider the dilemma a shareholder faces when the stock market undervalues her shares. The shareholder may believe the market will eventually come to correctly price the company's stock. But when? The market may not correct itself until after she has sold her shares.

In 1960, an average stock traded on the New York Stock Exchange would be held for eight years. Today, with the low cost and ease of trading frequently, as well as the deluge of information that floods the market and drives reactionary trades, the average holding time for a stock is just a few months. About 70 percent of all trades of U.S. stocks today are being conducted by "hyperspeed" traders, many of whom hold stocks for only a few seconds. Heeding the future as CEO of a publicly traded company is often at odds with the desire to please the new breed of shareholder. It's yet another way in which information usurps foresight.

Businesses, investment firms, and government agencies are all failing to correct for the myopia of individual shareholders. But this is a choice, not an inevitability. And there are ways to change course.

Andrew Haldane of the Bank of England suggests that investors need to be tied to the mast like Odysseus. This could come in the form of investment firms that hold on to stock for decades, well beyond the mere years of Eagle Capital, or through government policies and tax schemes to require or encourage longer stock holding times.

One policy that has been proposed by a group of American business leaders from both political parties is to issue a financial transaction tax to reduce the speculative, high-frequency trading in the market in favor of longer holding times for stock. The group calls itself the American Prosperity Project, and it includes CEOs and chairmen of companies such as Royal Dutch Shell, Levi Strauss, and Pfizer. The UK, Hong Kong, and Singapore all have variations of such financial transaction taxes in place.

Other reforms by boards, or governments, could encourage more shareholders and CEOs to invest in long-term projects. One option would be to keep executives from cashing in on stock compensation for periods longer than seven years, so that they are personally motivated to look well beyond the next quarter.

I favor ideas to give more voting power or ownership rights to shareholders who invest for a long time. The default approach of the U.S. corporation of one vote per share puts those with short-term interests on equal footing with long-term investors. Some companies have found creative ways to change this dynamic. Google, upon its 2004 initial public offering, created a dual-tiered system for its stock. One type of stock, retained by longer-term shareholders and founders, had ten times the voting power of stock held by the public, including short-term traders, giving the latter less influence over business decisions. In 2015, when the company reorganized itself as Alphabet Inc., it created a three-tier structure, including an expanded pool of shareholders who have no voting rights at all. It also happens to be a company that has made substantial investments in long-term research, and whose stock, as of this writing, has consistently been one of the most valuable on the market. As an alternative, long-term stockholders could be given warrants to buy greater stakes in a company over time.

If companies don't want to take these steps themselves, governments can step in. Dominic Barton of McKinsey & Company has suggested a reform in the United States to assign voting rights of publicly traded stock according to the average length of time that an investor holds stock. This would make the stockholders who in general behave more like owners more powerful in company decisions than those who behave like speculators.

Eric Ries, the Silicon Valley entrepreneur, has been working to create a long-term stock exchange that would bring together several of these ideas. Companies that are traded on the exchange would have to agree to compensate their executives based not on earnings per share in a quarter or short-term stock price, but on longer-term measures. Shareholders' voting power would be determined by the length of time they held their shares, and companies would report on long-range investments in R&D in greater detail. While some technology company leaders have backed the idea, it remains to be seen whether the exchange can get a critical mass of investors and companies, and whether it can adhere to its long view in the process. But it's possible for companies to adopt such practices even without going to a new stock exchange.

It will take bold action on the part of organizations to put any of these reforms in place, and as I write this I remain skeptical that we'll see widespread change to capital markets anytime soon. But these kinds of shifts can come about unexpectedly—either in response to crisis or if the right leaders take charge. For now—in the absence of radical changes—organizations will have to use the tools at their disposal to reconcile the demands of the moment with what is good for the future.

THE GLITTER BOMB

Lights Along the Long Road

Consult the genius of the place in all;
That tells the waters or to rise, or fall.
—ALEXANDER POPE,
"Epistle to Burlington"

The sun broke into the Kansas sky like an egg yolk and, as it rose, began to simmer. Wes Jackson, a farmer and native son of these plains, gazed out his truck window at a river that had changed.

A quick-witted iconoclast, Jackson grew up digging his hands in the dirt. He escaped a life as an academic scientist in California to return to Kansas in the 1970s. He wanted to find ways for farmers to earn profits each year without destroying the land that will feed people in the future.

He has committed his life ever since to resolving a profound trade-off between what farmers want today and what is good for their farms—and for humanity—in the long run.

The Smoky Hill River slugs for 560 miles from the flats of Colorado east to where it meets the Republican, then empties into the Kansas River. A trail blazed in the late 1850s runs alongside the Smoky like a sidekick, once littered with the hand tools and bones of gold miners bound for the Colorado fields that gilded the city of Denver. Bandits lurked on the trail and robbed the foot travelers. Icy winters made travel arduous and sometimes deadly.

I visited Jackson at his nonprofit, the Land Institute, a seven-hundred-acre sprawl of golden fields, tall grasses, and makeshift offices hugging the banks of the Smoky Hill River in Salina, Kansas. Jackson knows all the ancient stories of this land, dating back to when the glaciers pressed the rugged landscape into prairies. He was born during the Dust Bowl of the 1930s, when farmers dug up parched earth, gouging grass-lands to plant more and more wheat amid a drought. They sent soil spool-ing up in storms that shrouded the sun and rained down through barn ceilings. Black dirt piled up like blizzard snow against the doors of home-steads. It charged the air with static electricity that shorted engines, stranding cars dead on roadsides. Prairie dust clouds even drifted from the Great Plains to the East Coast, where they cloaked the Statue of Liberty and the U.S. Capitol Building in brown fog.

The drive for quick wins fueled both the Dust Bowl and the early gold rushes. The first panners profited by beating others to the gold. But the dust farmers who plowed up the prairie to plant wheat drove down prices and ravaged the soil, sowing a catastrophe. Hundreds of thousands of people lost their farms, their food, and their living. Jackson, who had just turned eighty before I met him, in 2016, still keeps harvest records from his father's farm dating back to 1934. "The dust was blowing the day I was born," he told me.

Today, earth bleeds from millions of acres of Kansas wheat, corn, and sorghum fields, not into the sky as dust but into the Smoky Hill River and other waterways. Thick as the innards of a tobacco chewer's cup, the river catches the fertile soil from farms, swept off by torrential midwestern rains. The farmers' annual crops have shallow roots that cannot hold on to the topsoil, and so it leaches off the land. More runoff means farmers need more fertilizer, water, and energy each year to sow their seeds again. Jackson sees this problem not in isolation but across a wide span of time.

For him, the muddy river calls to mind a mistake made thousands of years ago in human history.

Homo sapiens have roamed the planet for more than a million years, but have been farming for only the most recent ten thousand. When people first started planting seeds in the Fertile Crescent, in the modern-day Middle East, they cultivated grains as annuals that grew each year and were then replanted, much like the marigolds and geraniums in our gardens. The seeds from annual crops were easier to collect and plant anew as early farmers moved from place to place. They chose annuals probably because they grew fast, in a single season, and were easier to breed for certain traits, such as retaining their seeds like corn does. The seeds could be saved to sow the next year in a new place. Perennials take longer to establish, and it is harder to breed them to prevent their seeds from shattering like the dispersing heads of dandelions. Shattering seeds are harder to collect and carry to the next place a nomad might settle.

The wild plants our hunter-gatherer ancestors foraged, however, were perennials, like those native to the grasslands of the Midwest for the centuries before the Dust Bowl. Perennial wild rice once sustained ancient Chinese civilizations. Most ancient grains are very different from those we eat today.

Today's monocultures of annual corn, soybean, rice, and wheat crops sprawl across Earth's landscapes and demand a lot of labor, resources, and luck to make them grow. A lack of crop diversity makes farms vulnerable to disease and pests, requiring an arsenal of potent insect and weed killers. Fertile topsoil gets washed away; the soil is tilled each year to plant again and is treated with artificial fertilizers. The annual cycle of tilling and sowing is a ritual befitting a culture that uses up resources and starts anew, replacing the old rather than investing in longevity. As Earth's climate warms, the risk of both drought and flood grows, and the specter of the Dust Bowl looms like a thundercloud.

Most farm businesses measure the value of their crops based on annual yield—how many tons of food they can get each year from planting the crop on an acre of land. Using only this metric, farmers would never see the dangers of severe drought until dust is already rising. The annual yield metric is not an arbitrary number, however, but a market imperative that determines a farm's survival.

The rock-or-hard-place choice that farmers face resembles those that arise in many other businesses and industries. What makes sense in the short run because it is rewarded by the marketplace, like fishing all the red snapper out of the ocean, is not what's good for the long run, because it destroys the fishery forever. Investors want quick profits, but that mind-set does not support a business to pursue, for instance, years of scientific discovery to yield a medical miracle.

Wes Jackson wants to see farms endure. And he wants to make foresight possible in light of competing demands for immediate profit. What matters to him is the amount of food farmers can get from a landscape not in one season, but over generations. He has tracked the world's soils as they erode, losing their ability to feed growing populations. He has made it his mission to look far out, into the distant future.

On a trip with students to the Konza Prairie in the '70s, Jackson had what he calls his epiphany. The ancient prairie, which had been in Kansas since the last Ice Age, had endured. The soil bore no trace of erosion and had never been tilled despite annual harvests of hay. Its perennial plants of myriad species had deep roots; they had withstood past droughts, floods, and even the Dust Bowl.

"We were ignoring the genius of the place," Jackson realized then and later recounted to me, paraphrasing Alexander Pope. "The wild is our own library—it's our Alexandria."

Prairies last. Jackson wondered whether their qualities—resilience to extreme weather, protection of fertile soils, persistence over generations— could somehow be brought to farming. With a small platoon of graduate students, family members, and groupies, Jackson started the Land Institute to carry out the research he hoped would make his idea practical.

What has long withstood the test of time may offer clues about what can endure in the future. The engineer Danny Hillis, in building his ten-thousand-year clock, has also looked to the past to shape his plans. In an age of the ephemeral tweet and devices doomed for obsolescence within a few years, Hillis is looking back at engineering feats that lasted over long spans of history. While the design ethic of the digital age is to create products with the fastest and best technology of the moment, Hillis believes that principle does not apply to making something work for millennia.

For his purpose, Hillis studied what some believe to be the world's oldest working clock, in Salisbury Cathedral in England. Historians estimate that it was constructed in the fourteenth century. The clock does not rely on its original verge-and-foliot mechanism; rather, it's been easy to modify over the years. After Galileo's discovery of pendulums as time-keepers and metronomes, the clock was retrofitted with an anchor and pendulum in the late seventeenth century. Hillis believes that what has made the clock's longevity possible is that it has the qualities of transparency and modularity: People could see how it worked without any special knowledge or training, because its mechanisms are simple, and they could take it apart and fix it or update it in pieces.

Hillis is applying these principles to the building of the ten-thousand-year clock. At one point, for instance, he and his fellow engineers considered using a metal alloy of nickel and titanium, nitinol, as part of the mechanism to calibrate the clock each year on the summer solstice. Nitinol looks like steel but is a special material, because even if it gets deformed by heat, it later returns to its original shape. But Hillis felt that nitinol was not transparent enough for future visitors to the clock, who might not know about these "magical" qualities and might try to replace it with a piece of steel. Instead, he and his fellow clock designers are building a double glass chamber, with the inner chamber holding a piece of titanium foil that serves as a mirror, reflecting the sunlight that enters the clock on the solstice back and forth until the inner chamber heats up and expands. The mechanism is more obvious, he believes.

The clock's parts might have been made with the semiconductor technology that Hillis has worked with for much of his career. But even if he could make such parts last, they might be too obscure to passersby in ninety-five hundred years who want to fix the clock. Instead, the clock's components and gears will all be mechanical and—unlike contemporary cars programmed with computers—it will be clear how they work to anyone who tinkers with them long enough. The chimes, set to ring in unique patterns designed by the composer Brian Eno, can be taken apart and fixed without affecting the clock's pendulum and time-keeping mechanisms, and vice versa.

Many artifacts discovered by modern humans from the ancient world, such as the Dead Sea Scrolls and the Egyptian pyramids, were preserved by virtue of being in a dry climate. That's why Hillis and Jeff Bezos chose a site in the Texas desert for their ten-thousand-year clock. They also observed that across history, stories have often outlasted physical objects, such as that of the Temple in Jerusalem, a site of worship described in the ancient books of Judaism and Christianity. Their collaborator Stewart Brand championed the idea of creating an organization for the clock, so that there could be a group of people to keep telling its story, shepherding its construction. It seems to me that the Long Now Foundation's purpose is akin to a family in passing heirlooms or stories from generation to generation.

Organizations that want to create enduring projects might also look to the lessons of living organisms that have survived thousands of years—some standing sentinel for the birth of human civilization.

The artist Rachel Sussman has chronicled the oldest living flora and fauna on the planet. She took photographs of a living five-thousand-year-old California bristlecone pine tree, fifteen-thousand-year-old volcanic sponges in Antarctica, and a half-million-year-old Siberian Actinobacteria. Sussman, in conversations with scientists who study long-enduring plants and animals, extracted secrets of their longevity for her astonishing book *The Oldest Living Things in the World*.

The Pando aspen colony in Utah, for example, is more than eighty thousand years old, and it has persisted by virtue of self-propagation—cloning itself—and by slow migration to fulfill its needs for water and nutrients from the soil. It even survived the volcanic winter spurred by the massive eruption seventy-five thousand years ago on Sumatra. Its strategy—making lots of copies of itself—is one echoed by digital archivist David Rosenthal, formerly a software engineer in Stanford University's library, and an expert on the preservation of the ephemeral digital artifacts and records of our era. Lots of copies dispersed to different environments and organizations, Rosenthal told me, is the only viable survival route for the ideas and records of the digital age, stored in technological formats and media that will fast become obsolete just as floppy disks and videocassettes already have. A singular repository, like the burned-down Library of Alexandria, runs the risk of being lost in a catastrophic event. Even the Internet Archive, a nonprofit project to back up much of the World Wide Web, is at risk, Rosenthal believes, because its two copies are stored within striking distance of the San Andreas Fault. A group of eleven Canadian organizations that are working together to archive the national government's information is one example of such a distributed network of clones.

Another secret of some of the oldest living things on Earth is slow growth. Sussman documents what are known as map lichens in Greenland, specimens at least three thousand years old that have grown one centimeter every hundred years—a hundred times slower than the pace of continental drift. Meanwhile, the oldest bristlecone pine trees devoted their resources to sustaining single branches or clusters of pine needles while the rest of their branches might appear dead—just an essential thriving part, rather than a flourishing whole.

Learning of this called to mind an exchange I had with a longtime master of the ancient Chinese and Japanese art of bonsai who has perfected the practice of making young trees look old, and cultivating miniature trees that endure past any person's life. Trees that last centuries, he said, don't waste their energy flowering—they put their focus on enduring. In other words, they don't try to be good at everything. There is a lesson, perhaps, for organizations whose resources can be devoted either to rapid growth or to sustaining themselves over time, but often not both.

Wes Jackson dug into the past, then retrofitted what he unearthed to meet farmers' needs today and in the future.

By domesticating wild perennials and crossbreeding them with existing annual crops, Jackson and his colleagues invented perennial grain crops. Scientists and farmers in the U.S. Midwest, China, and Africa have been trying to perfect the crops over the past decade.

Perennial grains, unlike annuals, burrow thick roots ten to twenty feet deep into the ground. Plants with such entrenched roots don't require much irrigation and they withstand drought better. Perennial roots clench the fertile topsoil like claws and keep it from washing away. This makes it possible for a rich soil microbiome to thrive that helps crops use nutrients more efficiently. A field of perennials does not need to be plowed each year, and so more carbon remains trapped in its soil instead of escaping into the atmosphere, where it contributes to warming the planet.

Jackson believes perennial grains can anchor a long-term mind-set for agriculture—using less water and fewer resources each year, while preserving soil and native landscapes so that farmers can grow more food across longer spans of time.

To meet farm businesses' immediate needs, however, perennial grains cannot be wholly like the demure prairie dwellers that inspired them. They have to borrow from what has made annual crops so appealing for ten thousand years. In other words, they have to bear a lot of seeds for annual harvest. To win over farmers, Jackson had to breed perennial crops with high annual yields.

When Jackson began this work, he was told it was impossible—that there was an inevitable trade-off between the amount of energy a plant puts into its roots, which protect the topsoil, and the amount it puts into its seeds, which we eat. Colleagues pointed out that grain farmers want to maximize the harvest index—the percentage of a plant's total weight that is harvestable each year. If you made the roots more like perennials, you'd lose the near-term gifts of annuals—their high yearly harvest yield. If you made the seeds more like annuals, you'd lose the long-run benefits of perennials but have more to harvest today.

Jackson would not give up so easily. He had faith that science would help him and his team figure it out. "There's Methodist in my madness," Jackson is fond of saying. His zeal for nature, and penchant for quoting the scripture of farmers and ecologists and poets, can make him sound like a secular deacon.

His defiance of conventional wisdom reminded me of an iconoclastic leader from a very different industry.

The widely admired and pugnacious head coach of the San Antonio Spurs, Gregg Popovich, has exercised foresight where his peers have been reckless with athletes' health. Pop, as he is known by professional basketball players and fans, is the ringleader of one of the most successful sports franchises in U.S. history. Since 1999, his team has won five NBA championship titles, and the Spurs have made it to the NBA playoffs for twenty-one consecutive years as of 2018.

Pop pioneered the practice of keeping star players out of games for rest to prevent later injuries. In a nationally televised prime-time game in 2012 against the Miami Heat, Pop rested four of his star players because, in his view, they had been playing too many back-to-back games on the road. For this, the NBA fined the team a quarter of a million dollars, and Pop was publicly criticized for depriving fans of the chance to watch their favorite players. The market imperative observed by teams from Chicago to Los Angeles to Cleveland was to play your stars during prime-time games without regard to the later consequences.

Todd Whitehead, a sportswriter for FiveThirtyEight, the data-driven news outfit started by election whiz Nate Silver, calculated in 2017 that Pop had rested healthy players for entire games twice as many times as any other coach or team. In recent years, however, the practice has caught on, as other NBA teams have recognized the wisdom of resting key players during the main season, even if it means having an imperfect record. The upside is that players are more likely to endure the playoffs at top fitness and without injury. Instances of resting healthy players are now on the rise. Whitehead notes that ambitious rest campaigns for star players in the regular season have correlated, on at least two recent occasions, with NBA championships in those same years for the teams.

In a sense, Pop gave the short-term speculators in his industry the virtual middle finger. And he got away with it because of his track record of winning, the allegiance of his players, and the admiration of other team owners and coaches. No sane fan, owner, or general manager of a sports team would call for the firing of a coach who delivers victories as consistently as Gregg Popovich. We need leaders bold enough to fight for foresight in more arenas—and to do it, they need to be respected enough to set the trend.

While reclaiming the ancient technology of the prairie, Jackson's team of scientists at the Land Institute and their collaborators have drawn on sophisticated computational and genetics tools sharpened over the last decade. DNA sequencing allows researchers to rapidly select from a field of thousands of plants the twenty that have the characteristics they want to keep breeding for the next generation. The researchers keep an archive, a seed vault of each generation of grain that they breed, so they won't permanently lose important qualities of the plants along the way. They have also taken advantage of the fact that perennials have a longer growing season, and with added time each year can grow deep roots *and* produce large seed yields for harvest.

To get perennial grains into production, Jackson also had to figure out how to overcome farmers' aversion to taking risks on unknown crops, and their immediate fears of not having buyers for their product. Researchers from the Land Institute and the University of Minnesota have brokered deals for twenty farmers to plant fields with a perennial grain that resembles wheat. They persuaded the farmers by securing buyers willing to pay a premium for the grain, including an offshoot of the apparel company Patagonia, which is using the perennial grain to brew a beer, Long Root Ale, that it sells at stores on the West Coast. General Mills plans to make cereal with the new grain. In China, farmers have planted perennial rice on thousands of acres of paddies.

The perennial crop experiment is ongoing, as other scientists now carry out Jackson's vision. (He retired from running the Land Institute in 2016 to create a college curriculum that he thinks will encourage stewardship of the planet.) Jackson's scientific acolytes are now studying how various perennial crops, including silphium, a substitute for sunflowers in producing oil, withstand high temperatures and drought. They want to offer farmers an option that is better suited for a warmer climate.

After I had followed Jackson around fields for two days, peppering him with questions, we returned to his office, which, after a fire, had been rebuilt from reclaimed telephone poles. A black-and-white photograph of a relative's ranch in South Dakota, where Jackson spent a summer at age sixteen tracking eagles with a Sioux Indian teenager, hung on the wall next to maps and shelves cluttered with papers and tomes.

The small foothold of perennial crops teaches the art of the possible, though it remains to be seen how far it can spread into industrial agriculture. Reckless decisions still reap rewards today for many industrial farms, and the proverbial chickens have not come home to roost. I find the underlying strategy a promising one, however, regardless of whether perennial crops go mainstream. Jackson took an intractable trade-off of the short and long runs and invented a way around it.

Jackson's approach was to make what lasts over time pay in the short run—to lure businesses down the path to what's better for them and society in the future. Boosting annual yields, securing immediate buyers willing to pay premiums, and building the research case for how perennial grains thrive better amid immediate threats are all ways to make it easier for farmers to convert to perennial crops in the present and stay on course. This reminded me of the lotteries run by credit unions, described in chapter 3, to entice people to save now for their future. It also reminded me of glitter-bombing my friend when he ran his first marathon, meeting him at key mile markers on the course to shower him with glitter and keep his spirits up for the long race.

The tactic can work even on a grand scale. The civilization on the Pacific island of New Guinea survived seven thousand years past the onset of farming despite periods of severe deforestation and climate change, in part because people wanted to plant fast-growing casuarina trees. The trees help the soil retain nitrogen and carbon, making it more fertile and speeding up the time it takes for fallow land to replenish itself and serve anew as a farm or garden. Jared Diamond notes that the islanders, like today's farmers planting perennial grains, could see short-run rewards from planting the casuarina trees, which in groves offered timber and fuel, and in gardens offered beautiful shade and the pleasant sound of wind blowing through their branches and leaves. The trees had immediate allure, but also sustained a civilization across millennia.

In Japan in 2017, I met an executive from the Toyota Motor Corporation, one of the top-selling automakers in the world. The company has been around since 1933, when it was founded by Kiichiro Toyoda, the son of a textile loom inventor. (The spelling of the name was changed for the American market to Toyota.)

The executive, who had worked at Toyota for twenty-eight years, had a flair for quoting science fiction authors and had clearly fallen off the company's PR wagon. He was fond of calling the Toyota Prius the "first car to be powered by guilt." I asked him how the company managed to invest in cars with new technology behind them, including hybrid vehicles and hydrogen cars, when Toyota did not often recoup its investment in the first generation of those lines that came to market. This was the case with the Prius. It often took a few iterations for a car with new technology to pay off. I wondered if the stereotype of Japanese people as long-range thinkers, which thickly coats the myth of the company's founding, might have some truth to it.

But the executive told me the decision to invest in these long-run projects had a more practical, short-term benefit. Along the way, the company borrows insights that arise from long-term research-and-development projects—new materials or processes they develop, such as how to make a car's outer shell more rigid—and uses them in current top-selling models to either improve performance or cut manufacturing costs. In other words, they found a way to create immediate rewards that made their sacrifices for the sake of future products seem worth it now to company leaders and investors. They exported the knowledge to yield fast results, while plugging away more slowly at making higher-risk products that might, like the Prius, eventually become popular brands for a niche market. In Japan, where Toyota is a jewel in the crown of patriotic industrial pride, about half the cars the company now sells are hybrids.

Like wedding savings to playing the lottery or breeding perennial crops with high annual yields, the tactic is to create short-term wins that lure the company down the path toward something that has future pay-offs. A similar method has been used with success by Seva Mandir, a nonprofit in Rajasthan, India, that delivers childhood vaccines in areas where families often do not immunize children. Researchers at the Abdul Latif Jameel Poverty Action Lab at MIT found that when the organization gave out free small bags of lentils to parents, they brought their kids back for vaccinations at far higher rates than those not offered lentils. The tactic even saved money, because more kids got immunized at camps giving out free lentils, and so those camps did more vaccinations with the same number of workers as the less busy camps.

It is an open question whether the glitter approach could work to lure investors to look beyond immediate profit.

Andrew Lo, a professor at MIT's Sloan School of Management who previously founded a hedge fund, believes that such a tactic might work for biomedical research, which requires investors to have both patience and a high tolerance for risk.

In the 2000s, Lo's mother was diagnosed with lung cancer. He met with the leaders of a biotechnology company that had an experimental drug he believed might help her. It did not. Lo lost his mom and, within a span of five years, lost five other close friends and family members to cancer.

It has been more than forty years since Richard Nixon declared a war on cancer. While certain cancers have proven treatable in the years since, many still have no cure or treatment. Chemotherapy and radiation work in some cases, though they also cause considerable suffering. Very few cancers have treatments that allow patients to experience minimal side effects.

Lo asked the chief scientific officer of the company he visited whether financing had any influence on the company's scientific agenda of seeking treatments for cancer. The man's response, Lo says, was that funding— not the scientific questions or needs of patients—was *driving* the company's agenda. To Lo, that seemed completely backward. As an expert in financial engineering, he decided to channel his anger and sadness into finding better ways to fund biomedical research and drug development.

Lo soon observed what is widespread knowledge among experts in medical research: Finding new therapies for cancer and other diseases takes a lot of investment and a long time to pay off—if it pays off at all. The kind of research that has eventually yielded cures to particular cancers is expensive and high risk, with failure far more common than success. It can require between ten and twenty years and cost more than $2 billion to take a drug from its initial discovery to being marketed and sold as a treatment. Only about 7 percent of potential cancer drugs make it from the earliest stage of research to eventual approval for use by regulators. Along the way there are many possibilities for failure as therapies get tested in larger and larger populations of people.

When a cancer therapy, or any other treatment for a recalcitrant disease, does pan out, of course, the payoff can be great—in terms of both profit and the lives of patients and their families. But many more drugs fail, and eat up years of time and hundreds of millions of investment dollars. As in soccer, lots of shots on goal are needed for the few that make it in. Most investors would prefer to put their money either in sure bets with low risk over time, or in investments that have quicker potential for high payoff, such as technology start-ups. As a result, not enough research funding goes toward curing the major maladies of our time.

Lo came up with an idea that he thinks could address the problem. He advocates that biomedical and financial experts come together to form megafunds that combine the financing of long-term projects seeking cures for cancer with investments that yield short-term profits, such as buying stakes in the royalties from drugs that are about to go on the market. The latter tend to be funded by investors who are willing to put up cash only for a sure thing, getting a stream of the royalties from a patented drug that has already crossed many hurdles. The former, meanwhile, tend to be funded by a much smaller pool of venture capital run by risk-tolerant experts who have a detailed understanding of a biotech company's scientific rationale for its technology, enough to make an educated bet on its chances of creating a viable treatment.

The amount of capital going to long-term projects has been constrained by a shortage of both expertise and patience among most investors. The research has become more complex and riskier in recent years. At the same time, technology start-ups that quickly get to proof of concept and can have explosive returns have become more attractive options for venture capitalists.

Lo's idea is to combine these kinds of investments to create a diverse portfolio where there is some high-risk and long-term investment and some low-risk and short-term payoff. The long-run investment would go to projects that take different approaches to treating cancer—such as blocking angiogenesis of tumors, stem cell therapy, and immunotherapy—rather than betting on one particular approach. By packaging high- and low-risk investments together, Lo estimates that a great deal of risk can be taken out of the entire enterprise of funding biotechnology research and a lot more capital can be applied to curing disease. His models show that a $5 billion to $15 billion megafund, if carefully managed to invest in diverse projects whose successes are independent of one another, could yield attractive returns for equity and bondholders.

A major megafund has yet to be launched, but Lo's idea has recently been getting some traction. UBS, a global bank based in Switzerland, has launched a smaller-scale Oncology Impact Fund of $470 million. Separately, Lo has been involved in the launch of private holding companies that put his investment approach into practice by investing in the early-stage development of drugs for rare genetic diseases, Alzheimer's, and breast cancer. The next few years will be a test of whether his idea can work in practice.

A significant barrier to creating an effective megafund is that it requires broad and deep enough expertise in biotechnology as well as a sophisticated understanding of financial engineering. There's also a danger that Lo's concept of creating securities—in other words, issuing debt as part of the pool of investment capital—could mask the risk from the view of investors. They bear a resemblance at least in structure to the same mortgage-backed securities that played a key role in the 2008 financial crisis. Lo's research-backed obligations would need to be rated credibly, according to their actual risk, which he readily admits. Moreover, the investment approach would need to be carefully managed over time in order to avoid replicating the mistakes of existing pharmaceutical companies and investors that seek only low-risk products rather than breakthrough therapies for intractable diseases.

I t might not be possible to create short-term rewards, or glitter bombs on long roads, for everything that an organization needs to do over time. But it might be possible in some instances to create an enticing fantasy of the future that motivates people in the present. This is, in fact, what Timothy Flacke did with prize-linked savings. Not all the people who save through such programs win the lottery, but most make wise decisions about their own future because the prospect of winning the lottery—however unlikely—holds sway in the present.

It is the illusion of what we might become that draws people to Las Vegas, and to gamble away our money at the roulette wheel or slot machines. From the tarmac at McCarran International Airport, you can see the casinos of Las Vegas Boulevard rise in the distance—like mountains beckoning conquest. Replicas of New York's Empire State and Chrysler buildings seem to both mock and pay tribute to the originals. The Mandalay Bay resort shimmers in the desert sun like a sequined dress. Just to glimpse that city is to believe in the irrational possibility of remaking yourself, of getting rich quick, of getting away with transgression.

Just as the subculture of professional poker holds insights about resisting instant gratification, there is something to be learned from the very design of Las Vegas and its casinos, from its conjuring of an illusion—like a magician pulling a rabbit out of a hat.

Historic feats of human ingenuity have arisen from a strategy that resembles the Las Vegas illusion.

In 1919, a French-American hotel owner named Raymond Orteig put up a prize to encourage someone to accomplish the imagined feat of flying alone nonstop across the Atlantic Ocean, between New York and Paris. No one had made a solo transatlantic flight. There were multiple failed attempts over the course of eight years, including deadly crashes, in pursuit of the prize. The $25,000 Orteig Prize set forth a challenge that had yet to be overcome, as if erecting a mountain to capture the imagination of adventuresome climbers. None knew if they would succeed, yet they tried anyway, at their own risk and expense.

In 1927, Charles Lindbergh made the successful solo flight in the *Spirit of St. Louis* and won the prize. He became an instant global hero, and shortly thereafter, transatlantic air travel became a booming business.

The use of prize competitions to spur collective imagination is even older than that, however. In 1714, Isaac Newton and the British Crown put up the Longitude Prize—a challenge to improve naval navigation with a prize purse of £20,000. Ship captains used the sun or the North Star to calculate their latitude, but did not have a way of reliably calculating longitude at sea—that is, their exact position east or west of where they started. Ships would frequently run aground or crash into rock formations, sending crews overboard.

To the surprise of the era's scientific establishment, a clockmaker name John Harrison solved the problem by inventing a prototype of what's now known as the marine chronometer, which precisely calculates distance as a function of time passing. Newton had expected an astronomer tracking the celestial bodies to win. Harrison spent four decades coming up with various prototypes, expending far more than what made financial sense for him at any given moment. (He was awarded modest incremental sums by the British Crown, something like getting showered with glitter on a long race.)

In the late eighteenth century, Napoleon's government offered a 12,000-franc prize to anyone who could invent a way to preserve food for military troops traveling across barren lands, which yielded the techniques of modern food canning. Nicolas Appert, the confectioner who won the prize, spent fourteen years coming up with his solution—boiling food and sealing it in airtight champagne bottles.

Prize competitions have been resurging in popularity over the past couple of decades as a tool of philanthropies, businesses, and government agencies to focus attention and creativity on formidable tasks. As with glimpsing Las Vegas from the airstrip, prizes spur people to imagine

themselves as winners. Large cash awards and the promise of public cel-
ebration serve as motivation to compete. The $10 million Ansari X Prize,
for example, called for a group to put a manned spaceship into low-earth
orbit with private (not government) financing. The idea of private space
tourism imagined by the prize sponsors had long been seen as too
risky and dangerous, but the prize captivated people and helped them
envision what it might be like to win. It took eight years for a team to
succeed, during which twenty-six teams from seven countries competed.
The launch of SpaceShipOne by the U.S. team Mojave Aerospace Ventures in 2004 gave rise to the multibillion-dollar private spaceflight
industry. The prize inspired the inventors to imagine what might be
possible.

That same year, the Defense Advanced Research Projects Agency of
the U.S. government, also known as DARPA, held another kind of competition: a race of self-driving vehicles across the Mojave Desert with a
$1 million prize for the winner. DARPA is the experimental research arm
of the Pentagon, and had posed this challenge to call forth new technologies for autonomous vehicles that could be deployed in ground combat.
The competition motivated inventors to start tinkering and brought together a diverse group of technologists. Sebastian Thrun, who went on to
lead Google's successful effort to create a self-driving car, said this in
2017 of the prize competition: "None of what is happening in self-driving
today would have happened without the original challenge—it created a
new community."

Prizes might be used more widely as a kind of bait to lure businesses,
governments, and philanthropies to invest in research and invention for
which there is no market demand right now but that offer potential long-
run gain for an industry or for society. Organizations can use them inter-
nally to motivate employees to work on problems or externally to bring
fresh thinking to old problems. Some have tried to use prizes, for example, to bring solar power to rural areas or to create tools to diagnose diseases that afflict the global poor.

Prizes are not so useful for problems that are too complex or that are unlikely to have high financial returns even in the long run. And they cannot lead us toward what lies outside our imagination—the happy accidents that arise from a discovery. For that, we have to keep indulging our curiosity about the unknown.

PART THREE

Communities
and Society

7

HELL OR HIGH WATER

The Politics of Precaution

When Moctezuma went to meet them at
Huitzillan, he bestowed gifts on Cortés; he
gave him flowers, he put necklaces on him;
he hung garlands around him and put
wreaths on his head.

—BERNARDINO DE SAHAGÚN,
*on the Aztec emperor's 1519 encounter with
conquistador Hernán Cortés*

E ven giants stumble and fall.

From the Maya of Mesoamerica to the colonies of Easter Island
to the ancient Roman Empire, civilizations across history have fallen
from lofty heights. Echoing the patterns of self-sabotage that we experience in our lives and witness in organizations, entire societies have failed
to act on the warning signs of their demise until it was too late.

Failure of this kind is not inevitable, however. The societies of the
past that have survived in the face of existential threats have done so
because of shared values, wise community practices, and deliberate government decisions, Jared Diamond contends. In studying civilizations
across millennia that thrived where others failed—from the Inuit communities of Greenland to the Tokugawa society of Japan to the long-
standing inhabitants of the Pacific island of New Guinea—Diamond

found that each adopted cultural practices and strong institutions that helped them endure. Many created ways of passing resources and knowledge across generations.

In today's modern societies, widespread practices prevent us from heeding warning signs. But we can still choose wisdom over recklessness.

They weren't holding a gun at your head, were they?" asked the plaintiff's attorney, with a hint of sarcasm, as he cross-examined the witness.

"Not literally," replied Kit Smith.

A gun barrel at her temple, in fact, seemed an apt metaphor for her sensation during the days in question—thirteen years before she was called as a witness in the trial. At that time, she had served as the chair of the council of Richland County, home to South Carolina's capital city, Columbia. The body of eleven elected officials was charged with overseeing land-use decisions in the county. In 1999, a group of influential investors had asked her and the council to quickly green-light a real estate development. The investors, who named their company Columbia Venture, proposed to build a billion-dollar "city within a city" on a vast tract of land just south of the capital. The site lay on the banks of the Congaree River, behind earthen levees, in a historic floodplain. In the days before they closed the deal on the land, the investors sought Richland County's assurance that it would take on the liability if the levees ever failed.

The city of Columbia lies at the confluence of two rivers, the Broad and the Saluda, which come together to form the Congaree River. Swamps dense with bald cypress, water tupelo, ash, and oak trees shroud the Congaree's shoreline as it wends south and east from the city, to its eponymous national park. Escaped slaves once hid in maroon colonies in the river's floodplain, cloaked by the tangled forest. During Prohibition, moonshiners tucked their stills and liquor stashes along the Congaree. The landscape remains composed of large tracts: former slaveholding plantations that have since become privately held hunting clubs and estates.

Between Columbia and Congaree National Park, straddling Interstate 77, lies the forty-four-hundred-acre site where Columbia Venture planned to build the project it dubbed Green Diamond. A tributary to the Congaree, Gills Creek, with dams staggered along its length, runs from neighboring Fort Jackson across the property.

For generations, the tract belonged to a farming family. At the time of the proposed development, some of the land was being cultivated, and the rest was largely undisturbed except for the occasional hunter, spotted on early mornings toting a shotgun and perhaps a flask of whiskey. Man-made agricultural levees held back the river, but still, the fields, like a vast birdbath, filled often with standing water. Near the property, also in the floodplain, was a city wastewater treatment plant and a private Episcopalian grade school whose gymnasium and cafeteria were elevated on several feet of concrete blocks. On the property proposed for Green Diamond, the manholes did not lie flush to the earth but instead were on pylons rising more than ten feet above the soggy ground.

Green Diamond seemed inevitable at first. In the summer of 1998, at the time the project was being conceived, Columbia and Richland County officials were eager for new development. They welcomed the prospect of more jobs and tax revenue to the area. Many hoped that Green Diamond, abutting the highway, would bring businesses, better housing, and infrastructure to outlying rural neighborhoods of the county—home to mostly poor black residents.

The early proposal for Green Diamond seemed to Kit Smith like a boon for the region. The developers promised twenty thousand jobs and an expansive commercial and residential community that would include a technology park, golf course, outlet mall, retirement village, medical facilities, restaurants and hotels, and hundreds of single-family and multifamily homes. The housing crisis was still a decade away. Hurricane

Katrina had not yet shown the world that an entire city could be devastated, and thousands could drown, when levees failed.

Columbia Venture enlisted big political guns to lobby for Green Diamond across party lines. It hired the former chairman of the Democratic Party's national committee. To influence the mapping of the floodplain, it would eventually meet with the U.S. Federal Emergency Management Agency's general counsel and its future director, Mike Brown, an appointee of President George W. Bush.

But early in 1999, in the days when the developers were pressuring Kit Smith and her fellow council members to give the project their quick blessing, Smith began to have misgivings. She had only recently learned about their request for the county to take on the responsibility of securing the levees—and of their request for $80 million in revenue bonds. Columbia was not a coastal city accustomed to developing in flood zones or fixing levees. Yet Smith felt pressured to just say yes. The deep-pocketed developers, her fellow council members, and the public's support had created mounting social pressure to make this development happen as quickly as possible.

Green Diamond posed serious danger to the community. Before Columbia Venture set out to buy the vast parcel of land along the Congaree River, the government agency that surveys floodplains across the country for insurance and planning published a new draft map of the region. The map showed 70 percent of the future Green Diamond property in what's known as the floodway—the most dangerous part of a floodplain, where waters reach their greatest depths and will rush at their highest velocity during a flood. During severe storms, people caught in floodways in coastal communities like Houston and New Orleans have seen water inundate their homes, submerge their cars, and drown their relatives and neighbors.

Every five years, FEMA, the Federal Emergency Management Agency, revises its national maps to reflect new information about flood risks. As time goes on, the likelihood of and damages from floods tend to rise due to urban sprawl. More houses, roads, and developments leave less room for rivers to spread through their natural floodplains and, like rocks dropped into a glass of water, raise the level of floodwaters during storms.

For communities, heeding national flood maps in making decisions about local development is voluntary. But it's a requirement if the community wants to participate in the National Flood Insurance Program—which underwrites and subsidizes insurance for homeowners—and for a community to be eligible for federal disaster relief when it needs it. As of 2016, the National Flood Insurance Program owed more than $20 billion to the U.S. Treasury, because it had paid out more to victims of disasters, including Hurricane Katrina on the Gulf Coast and Hurricane Sandy in the Northeast, than it earned in premiums. By the time Hurricanes Harvey, Irma, and Maria hit in 2017, the program had reached its borrowing cap to pay out claims.

One house in Spring, Texas, documented by journalist Mary Williams Walsh, had cost the flood program—and U.S. taxpayers—$912,732 to repair nineteen times, even though it was worth only $42,024 in 2017. It is but one of tens of thousands of "severe repetitive loss properties" that get flooded over and over again. When Congress tried to raise flood insurance premium rates in 2012 to reflect the growing costs of disaster relief, many coastal citizens vehemently objected. And so the program's debt kept growing as people around the country continued to build more homes and businesses in high-risk floodplains. Imagine if the government similarly discouraged people from wearing seat belts when driving on highways.

Not every policy is so reckless, however. FEMA asks communities not to allow any building in floodways that will lead to a rise in flood levels. More than thirteen hundred communities around the country have enacted even more cautious programs of their own volition. Tulsa, Oklahoma, has cleared nearly a thousand buildings from its floodplains in the past thirty years via a city effort to buy back homes and businesses after destructive floods in the 1980s. King County, Washington, home to Seattle, keeps more than a hundred thousand acres of floodplain as natural open space. FEMA gives residents of such communities lower rates on their flood insurance to encourage more communities to prepare for flooding in a similar fashion.

It often takes the experience of a disaster to spur a community to take such action; deadly flash floods in the summer of 1997 in Fort Collins, Colorado, for example, led the city to prohibit any new development in floodways.

In 1994, Richland County, South Carolina—not on the heels of disaster—had taken precautionary measures. Like Fort Collins, the county had, through a local storm-water ordinance, prohibited building in the worst hazard zones, the designated floodways. This measure later lowered flood insurance premiums for people living in the county. Local officials I spoke to believed it also protected residents and first responders from being dispatched to rescue people during storms.

The powerful developer behind Green Diamond wanted to get around the problem posed by the fact that 70 percent of the property was in a floodway, where local regulations would not let it build. The company asked Richland County to weaken or waive its rules. In the past, the county had granted permission to people requesting to build a boat dock or pier in a floodway, for example, but it had never received a request to put millions of dollars' worth of commercial property and homes for thousands of people in a floodway. The developers' other tactic was to lobby FEMA to change the map so it would no longer show most of the property lying in the highest-risk part of the floodplain.

Kit Smith had a bad feeling in the days she was being asked to give the go-ahead to the development. So she called the South Carolina state floodplain coordinator, Lisa Holland. Holland, whose surname is now Sharrard, told Smith that building homes and an elderly residential community on that site could put thousands of people in danger, and that it could be devastatingly expensive for the county if the levees failed.

Government flood maps are drawn based on the history of floods, taking into account the growth of urban infrastructure and the shifting topography of rivers over time. The focus of FEMA's concern is the hundred-year floodplain, an area that in any given year is supposed to have a 1 percent chance of flooding.

In October 1929, just before the stock market crash that ushered in the Great Depression, two tropical storms hit South Carolina. Records show that the Congaree River reached a flood stage of 152 feet in the city of Columbia, washing out bridges on its tributaries and closing down highways, mills, and hydropower plants for days.

The risk of a disastrous flood and devastation in the Congaree floodplain would increase after Green Diamond was built, Sharrard told Kit Smith. The development would add more buildings and pave over more of the natural landscape of the floodplain. And the warming climate was expected to bring heavier rains and more intense flooding to the southeastern United States. The flood map did not reflect that, because it was based on historic and not future flood risks. Everyone was underestimating the danger of building in that floodplain, Sharrard believed.

Caution pays off in the long run for communities that face a high risk of natural hazards. For example, Wharton economist Howard Kunreuther calculated in 2009 that if the states of Florida, New York, South Carolina, and Texas simply updated residential building codes by applying current standards to old properties, they could save tens of billions of dollars in hurricane damages. A report commissioned in 2017 by the U.S. government showed that for every dollar people, communities, and governments spend preparing for earthquakes, floods, and hurricanes, six dollars is saved in responding to the disaster and rebuilding after the fact. Many disaster experts believe that communities and societies now actually save far more—as much as eleven dollars for every dollar spent—when they take the precaution to build wisely and protect residents and property, because of the rising tolls from disasters in a warming climate.

Unfortunately, it's not a clear-eyed calculus of present and future costs that guides most community decisions. There are immediate rewards, and distractions, that motivate political leaders and even entire societies to disregard the future.

The way communities and societies measure progress, for one, often favors shortsighted decisions—and masks the devastation wrought by disasters.

Since the mid–twentieth century, gross domestic product has been the dominant way to gauge the well-being of a country. Yet it is a fallacy that GDP measures a nation's true welfare.

It is an especially poor proxy when a country is destroying its natural resources and barreling toward civil unrest while its economy temporarily grows. In this way, GDP growth can actually hide that a country is on a reckless path. Nobel Prize–winning economists Joseph Stiglitz and Amartya Sen offer the example of a poor country that awards a mining lease to a company without securing enough royalties or putting into place laws to prevent harm to human health from air and water pollution. GDP may rise while the well-being of the country and its people declines. Traffic jams might increase GDP because of greater use of gasoline, and at the same time degrade the quality of people's lives by raising their stress levels and endangering their health. Destructive earthquakes or cyclones can actually boost a country's GDP in the immediate aftermath, because of the spending for recovery, even if they wreak permanent humanitarian and economic damage. This temporary jolt has been documented after natural disasters around the world.

Sen and Stiglitz note that even GDP per capita, the common measure of progress for people within a country, obscures inequality. From 1999 to 2008, for example, GDP per capita was rising in the United States even though most individuals over that period saw a decline in their income, adjusted for inflation. In hindsight, we know that inequality was growing on the eve of the financial crisis, even as income overall was growing.

Societies of the past have imploded rapidly, Jared Diamond writes, just after reaching their peak power and size. Why were people in those civilizations taken by surprise? A key reason is that the signs of imminent decline—such as critical resources being depleted—were masked by short-term fluctuations in resource levels. This is similar to how, for a few decades, the rise of superbugs was masked by the temporary solution of new antibiotics being invented, and to how cabdrivers are deluded by meeting daily targets even when missing their yearly goals. It's a problem of fixation on the dashboard. Diamond writes about how the former inhabitants of Easter Island likely did not notice from year to year the long-term trend of deforestation that destroyed their thriving civilization by the eighteenth century. Annual forest cover change, he argued, would have been barely detectable, the larger trend belied by the slow-growing saplings sprouting on denuded lands. Entire communities and societies can be distracted by single gauges on the dashboard.

Tyler Cowen, an economist at George Mason University, points out that GDP fails to capture critical aspects of countries' well-being, including health, environmental amenities and resources, leisure time, and the work done within households that is not bought or sold, such as caring for the elderly and children. It captures only the goods and services that are bought and sold each year.

Cowen advocates a "wealth plus" measure that would replace GDP, and reflect all that contributes to human welfare and a society's well-being. It's not so straightforward, however, to measure caregiving, equitable access to wealth, resource conservation, or leisure time as it is to count the number of manufactured widgets sold. The difficulty of pinning down such numbers, in large part, has prevented a saner metric from supplanting GDP. From Sen and Stiglitz's point of view, any singular target is unlikely to capture what ought to be measured in a society. I tend to agree that we need to use multiple metrics—and even find ways to look beyond numeric targets altogether to ask deeper questions about where a society is headed.

Another reason communities and societies fail to exercise adequate foresight about future disasters is that what pays off in the future often does not pay off politically today. Candidates and elected officials get credit for responding to crises after they happen, not for the hidden act of averting a crisis altogether. The September 11 World Trade Center attack made New York City mayor Rudy Giuliani *Time* magazine's "Person of the Year" and launched him into the national spotlight. Had he been somehow able to prevent an attack in the first place, it's likely fewer people would have noticed.

Part of the reason for this, of course, is that the horrors of the September 11 attacks in the United States were not fathomable to most people until after they had happened. Leaders of communities and societies, and voters, suffer from the same failures of imagination that all of us do in our lives, whether we're trying to imagine old age or our next camping trip. While we cannot anticipate everything, some risks are obvious and should be on our radar.

Failures of imagination have recurred like a nightmare in my work. In late 2014, I joined a group of doctors, scientists, and policy experts who met in Boston during the Ebola epidemic. It was the peak of the American public's fears about the deadly virus, which was spreading beyond West Africa, with cases that had just cropped up in Texas and New Jersey. The epidemic eventually killed more than ten thousand people worldwide, most of whom lived in Liberia, Sierra Leone, and Guinea.

The group's charge was to come up with ways to respond to the crisis to pass along to the White House. But I could not help feeling frustrated that we'd gotten into this mess in the first place. It was not until eight months after the outbreak emerged in West Africa that the World Health Organization deemed it a global emergency—a designation that triggers nations around the world to take aggressive action to stop an epidemic. During those months, some of the worst damage wrought by the Ebola epidemic might have been prevented by a stronger response to contain and treat the cases that had emerged. Only after nearly a thousand people had died and the epidemic was fast spreading across borders did the WHO name it a global emergency.

E-mails later published by the Associated Press revealed that officials knew of the potential danger and scope of the epidemic months before the designation, and were warned of its scale by the nonprofit Doctors Without Borders, which provides medical treatment to people in war-torn and remote regions of the world. The World Health Organization's leaders, however, were worried about declaring the emergency because of possible damage to the economies of the countries at the epicenter of the outbreak. In 2015, Michael Osterholm, an infectious disease epidemiologist at the University of Minnesota, compared this excuse to not calling the fire department when several houses are on fire "because you're afraid the fire trucks will create a disturbance in the neighborhood."

The human toll and economic devastation in West African countries turned out to be far worse than any intervention would have been. Humanitarian aid pledges rose to billions of dollars, the Liberian economy faced near collapse, and airlines marked multimillion-dollar losses in the year of the outbreak. Thousands of people died tragically and needlessly. The outcome was not unforeseeable, however. As the outbreak was emerging, experts and world leaders armed with the tools of epidemiological research and historic outbreak records could have conceived of it. But the outcome lay outside what they actively imagined. Competing immediate concerns, however minor, infected their thoughts.

In 2020, an outbreak of a novel coronavirus in Wuhan, China, turned into a pandemic not because it was a surprise threat. Leaders in government, business, and media—particularly in Europe and the United States—underestimated its potential global reach and devastation and acted too late to contain the epidemic's spread. The U.S. government had also failed during the preceding years to sufficiently invest in preparing for such a pandemic, for example, by neglecting to build up its stockpile of medical equipment like surgical masks and ventilators.

The historian Barbara Tuchman has defined folly, the kind that has

led countries to fight losing wars and destroy triumphant empires, as a
society's failure to act on knowledge that leaders perceived at the time—
even when there were feasible alternatives and no single tyrant was in
power. When the Trojans accepted the wooden horse, when Moctezuma
sent gifts to Cortés, when America invaded Vietnam—the choices in
Tuchman's estimation were "marches of folly." Societies and leaders of
each era knew better but acted as if ignorant. To me, the responses to the
Ebola epidemic and the coronavirus pandemic fit that mold.

The lead investor and managing partner of Columbia Venture was a legendary South Carolina company known as Burroughs & Chapin. It
hails from coastal Horry County. Burroughs & Chapin built modern
Myrtle Beach, its sprawl of golf courses and beach resorts, strip malls and
amusement parks.

The transformation of Myrtle Beach from a rural backwater into a
resort town was the brainchild of a nineteenth-century patriarch, Franklin G. Burroughs. A native of the river town of Conway, which lies just
fifteen miles inland from the beach, Burroughs made a fortune harvesting the sap of pine trees for tar, pitch, and turpentine. While growing his
business, he realized that it was cheaper to buy land along the coast—
priced low in that era because it was too brackish for farming—than it
was to lease it for the trees. By the time of his death in 1897, he owned
most of the land that is now Myrtle Beach. Before he died, he shared
with his sons his unfulfilled vision to develop that land into a beautiful
coastal enclave.

The Burroughs sons created a land holding company at the turn of
the twentieth century and began to lease oceanfront lots to hotel and
resort developers. They sold land to future residents—for the whopping
price of $25 per lot. The family held on to aspects of their father's vision
as time passed, building public rights-of-way to the beach instead of selling all of the lots to residential property owners who could block access.
Nevertheless, the development mushroomed beyond F. G. Burroughs's

vision; today Myrtle Beach is a mishmash of go-kart race tracks, oversize vape shops, strip clubs, pastel-hued hotel towers, motorcycle bars, and posh private clubs. The traffic in the summer is stifling, and coastal wetlands have been paved over to accommodate the cars of surging seasonal populations.

Over time, the company that became Burroughs & Chapin earned the goodwill of the residents of Horry County, donating land when parishioners wanted to build a new Episcopalian church and when the city of Myrtle Beach decided to build an art museum. The family's roots reach deep and wide along this stretch of South Carolina's coast. To live in Conway, the county seat—a place where I have visited close friends over the course of twenty years—is to know a Burroughs personally and to know the family's good deeds by name.

Susan Hoffer McMillan is a local historian of Horry County and of the Burroughs family, a former journalist, and the author of six books on Myrtle Beach. She is also married to the great-grandson of Franklin G. Burroughs. (Her husband retired as the company's vice president and chief financial officer in 2000, just after Green Diamond was proposed.)

Original portraits of the deceased nineteenth-century patriarch and his wife hang on the wall of her spacious home in Conway, on a street lined with sprawling magnolia trees and live oaks dripping with Spanish moss. Her home was built by Franklin Burroughs's widow and her daughter in the early 1900s, and has remained in the family ever since. We sat in one of several expansive and gorgeously adorned living rooms, with colorfully patterned sofas, Tiffany lamps, and fishing floats all in a dazzling cobalt blue that glowed in the afternoon sun.

McMillan believes the Burroughs family company originally had a philosophy of slow growth, a pace that lasted for two generations. In her view, it was in the 1990s, under a new CEO, that the company started to grow aggressively and to look beyond Horry County to the state capital and to Nashville, places where it lacked a track record and had not earned the trust of the community. It had never before tried to develop a property with levees or to build a project outside its home turf. In the same period that Burroughs & Chapin took the lead role in Columbia Venture to propose Green Diamond, the company's pace of development skyrocketed, and so did its debt.

In many towns in South Carolina, everyone seems to know everyone—or at least they know someone in everyone's family. It is a deceptively small place, and at times during my visits the entire state of five million people has felt more intimate than the midwestern town of twenty thousand people where I grew up. As a visitor, you'll find people often know whom you met with earlier that day, and whom you are meeting with tomorrow. The small towns, in turn, feel more like extended families, with offbeat characters who are ridiculed but tolerated, and bitter feuds.

It would be hard to overstate the social pressure that local officials in South Carolina felt to speed ahead with the Green Diamond development. That pressure came along as a side dish to the influence of deep-pocketed developers with high-level political connections and the ability to fund local candidates' reelection campaigns. The climate was ripe for overlooking long-term risks in favor of immediate returns to developers who were unlikely to stick around for the long haul of whatever happened to the property. The community had become like a casino. In the professional and social circles where Kit Smith spent her time, she got the strong message to favor the urgent business at hand over the future.

"When you have that much money and influence coming in like a sandstorm, it's almost impossible to resist," Lisa Sharrard, the flood expert, told me. "The developers came in whistling Dixie, and almost everyone was falling in line."

Reckless decisions to develop land are often written off as a by-product of greed or ignorance. An unsightly strip mall or high-rise is sometimes, of course, partially the work of rash and money-grubbing politicians. But the law has played a more pivotal role than many people realize in collective myopia about land use. In the United States, this perversion of foresight by law dates back to the founding of our country.

After the U.S. Constitution was ratified by the states, Thomas Jefferson had suspicions about the growing powers of the federal government. Along with his fellow anti-Federalists, Jefferson pushed for the amendments that became the Bill of Rights. The Fifth Amendment, in what's called the Takings Clause, states: "Nor shall private property be taken for public use, without just compensation." The idea of a "taking," as the founders wrote about it, referred to the national government seizing private land from citizens to build military forts or roads. After the Civil War, new amendments to the Constitution were added that made the Bill of Rights, including the Takings Clause, extend to state and local government actions, not just those of the federal government.

In the early twentieth century, property owners began to invoke the Takings Clause to argue that the government needed to compensate them—not just when it seized land by eminent domain to build a highway or military outpost, but if regulations prevented them from getting the full economic benefit of their land. Lawsuits ensued, with developers and landowners demanding large settlements even in situations where the government was trying to serve the public interest—with zoning to protect public health or prohibitions against paving over wetlands that protected people from storms.

How did the idea of a "taking" morph to include laws that protect the public? The metaphorical take on the Takings Clause got traction with a 1922 U.S. Supreme Court case. At the time, a company called Pennsylvania Coal owned belowground mineral rights to many properties in northeast Pennsylvania. The state legislature, hoping to prevent nuisance to residents, passed a law to prevent mining under people's houses, city streets, and areas of public assembly like town squares. The coal company saw this as a deprivation of the value of its subterranean property, and argued that the Pennsylvania law was a taking. The U.S. Supreme Court ruled in the company's favor and invalidated the state law, letting the company mine under neighborhoods.

For nearly a century since the coal-mining case, cities and states around the country that might have passed laws to prevent reckless development have had to consider whether they might get sued by a landowner or developer to compensate for a taking. Developers have wielded the Pennsylvania Coal decision to fight national policy that curbs development and, in some instances, to intimidate local planners. Some communities that want to prohibit building in hazard zones fear a lawsuit that could drain their coffers—or worse, plunge them deep into debt if they don't zone carefully. Meanwhile, local officials also experience political pressure from home builders associations that affect their reelection prospects. Local leaders are encouraged to be reckless—to ignore the future risks of developments and approve them quickly to indulge the near-term rewards of more tax revenue, jobs and housing, and profits to developers and businesses that help keep them in elected positions. Even good intentions to exercise collective foresight can be little match for these forces.

In the latter half of the twentieth century and into the twenty-first, unwise developments have sprung up across the United States, raising the costs of extreme weather events like floods, tornadoes, and wildfires, because more and more people are living in harm's way. Since 1980, just 233 extreme weather events have caused more than $1.5 trillion in economic losses. The disasters have killed thousands of people. In 2016, which broke historical records as the warmest year on the planet, there were fifteen climate and extreme weather events in the United States whose costs exceeded $1 billion each. In 2017, the record high was repeated. Globally, extreme weather and climate disasters cost more than $250 billion each year; over the span of a few decades, millions die and billions are injured in these catastrophes.

It's even harder to persuade people to move out of harm's way than it is to prevent risky development. Over time, people put down taproots in communities, anchoring their sense of place, of belonging. We see people choosing to stay in their homes and neighborhoods in spite of knowledge that waters will rise, whether they are working-class families of Houston or wealthy people along the Mississippi Gulf Coast. Even when storms or fires or earthquakes come, it's difficult to persuade people to move to safer ground. Relief funds, ostensibly a tool of persuasion, are politically perilous to wield in times of disaster. Imagine a governor or mayor who tells people who have just lost all their family photos and furniture that they can't move back into their neighborhood. It's easier to ask people to make marginal improvements, such as elevating their homes or adding rain spouts to their garages, than it is to ask them to move out of harm's way.

The politics of precaution can paint a grim picture for communities seeking to make wise decisions. There is hope, however. In the end, Richland County showed how a community can wield its power to plan ahead.

Kit Smith is not the kind of person you'd expect to be shunned at cocktail parties. She hails from a prominent South Carolina family, and has the charm and connections that typically win people favor.

Kit grew up in Gaffney, a town of twelve thousand that's home to the Peachoid, a 135-foot water tower painted like the fruit, visible from Interstate 85 as you drive between Greenville and Charlotte, North Carolina.

She learned early in life to overcome what she sees as her natural introverted tendencies. She lost her first campaign, for the elected position of cheerleader in seventh grade at her football-crazed school, because she barely talked to anyone in the halls. The next year, she started smiling and getting to know her classmates, and won a spot on the squad.

After she had learned how to people-please, Kit frequently abandoned the pursuit in favor of her principles and, over time, even gained comfort with iconoclasm. She fought bullies, earning the nickname "Kit's Kittens" after wrestling with a volleyball player she thought was cheating at the game. As a senior, she slapped a six-foot-five basketball player who said something rude about her father. In 1978 in the heart of the conservative South, at age twenty-nine, she became the president of her local chapter of Planned Parenthood. Everywhere she went, even with her young son in tow, she imagined people whispering behind her back, "She hates babies."

People who know Kit Smith, whether they like her or not, tend to respect her. They call her congenial, clever, politically adroit, and feisty. Smith likes to be liked, but she has a reputation of sacrificing admiration for her beliefs. "I actually like attaboys," she confessed to me in her thick drawl. "But somehow I can do without 'em."

Her first foray into government was serving as research director for the medical affairs committee in the South Carolina Senate. After several years, she left for the private sector, where she worked in public relations. In the late 1980s, she felt the urge to get back into government. She ran for Richland County Council and won. In her first campaign, she got support from developers and home builders associations. They knew her husband was a banker and a member of the chamber of commerce, and they figured Smith was a safe bet to be pro-development. By the time of the Green Diamond proposal, Smith had stopped earning the favor of campaign donors because of her independence, yet she continued to win reelection. Her name and wealth made it possible to self-fund her campaigns.

As Smith grew more convinced by research that the Green Diamond project was a bad idea, she also grew isolated from her peers in her opposition. The real estate proposal began to metastasize into an acrimonious battle for the fate of the community. Smith went to the press to express her doubts. Attacks volleyed in both directions. Smith was painted as a backstabber by supporters of Green Diamond, a rich woman who just wanted to lace up her fancy hiking boots to walk along the river. Lobbyists for Green Diamond suffered public accusations from friends that they were selling out their community just to buy second homes.

Smith had never felt so hated before getting in the way of Green Diamond. Soon she was getting dirty looks at church and at her book club meetings. Fellow council members who were once allies attacked her for overlooking the needs of the black community, who had been her constituents before she was elected in a different district. The idea that she was deliberately depriving them of the opportunities promised by Green

Diamond stung. She began to spend a lot of nights without sleep, and at one point she wrote a manifesto calling Green Diamond the "zircon of greed." But she still didn't know whether she would be able to stop it.

Myopia is widely encouraged in modern democracy, particularly in the United States. Politicians face the immediate need to win elections—and to secure money for the arms race of financing their political campaigns—much like corporate leaders face the urgent task to boost quarterly earnings. Frequent election cycles require quick wins to prove progress to voters—regardless of the future consequences of reforms. Unfortunately, many of those who have the resources to self-fund campaigns use them to seek their own profit, rather than the ultimate benefit of all. Independence from campaign financiers can give leaders more latitude to govern with the future in mind, but it's not a guarantee that they will.

Of the forces that affect political folly, Barbara Tuchman writes, the "lust for power" reigns supreme. Leaders take their nations into losing battles and invite in enemies when accompanied by beautiful enough Trojan horses, even when they know better.

Every type of society and regime has suffered from such folly—despite knowledge at the time of the likely eventual consequences of their actions. Communists and capitalists, tyrants and democrats alike have been prone to making counterproductive decisions based on their immediate desires. This is Tuchman's account in her acclaimed book *The March of Folly*. Not even religious leaders are exempt from warped decision making, as she documents in her analysis of the Renaissance popes whose calamitous choices provoked the Protestant Reformation. We need not look far in our own time for examples of greed among elites that prevents wisdom about the future—whether it's oil companies' influence on pollution rules or corrupt politicians who line their families' coffers while holding office. The shortsightedness of those who seek power cannot be easily corrected without a populace that demands greater attention to future consequences on the part of its leaders.

One radical experiment in Africa meant to fight political corruption puts a fine point on the problem. Mo Ibrahim, a billionaire who built his fortune selling Africans mobile phones (which leapfrogged the spread of landlines on much of the continent), is the founder and benefactor of a large cash award for former heads of state. The Mo Ibrahim Prize rewards its recipients—outgoing political leaders of African nations—with $5 million over ten years after they leave office and $200,000 per year after that for their lifetime. In choosing a winner, the prize board evaluates whether a leader fought corruption and supported democratic reforms. It aims to counter politicians' short-term motives by offering pro-democracy leaders more than enough money to retire in style in exchange for creating enduring legacies—and for leaving office on time. Some years, no prize is awarded because there aren't qualified candidates. It seems that far too few politicians have proven worthy.

A major flaw with the prize design, in my view, is that it expects politicians to be able to value the money they'll get in the distant future over their present concerns. The prize might work better if it offered near-term rewards to counterbalance the politicians' immediate temptation to make decisions to get funding for reelection or please special interests. The problem with that, of course, would be that the prize would then effectively be a bribe. It would be hard—and possibly illegal—to create a scheme that has enough credibility and integrity to give politicians prize money now for what they do for the future. We do see campaign donations come from people who ask politicians to defend long-term interests like public health or conservation, but they can be paltry in comparison with the funding provided by companies and industries with short-term goals.

Better yet would be if governments around the world could undertake true campaign finance reform—limiting the amount of money donors can give to candidates and strengthening anticorruption laws—in order to separate politicians' decision making from the powerful forces that train focus on the present at the expense of the future. Reelection prospects should not be contingent on whether a leader can raise money from private funders, in part because any particular funder can seek short-term gain at the expense of the collective future.

Campaign finance reform at the national level has been flagging in the United States for more than a decade—and it faces new challenges given the Supreme Court decision in *Citizens United* in 2010. (In this ruling, the Court held that spending by corporations and unions to advertise on behalf of political candidates is protected speech, not to be curtailed by government regulation. It has unleashed torrents of corporate and special-interest funding aimed at swaying election outcomes.) *Citizens United* has brought us an era of American leadership and decision making more geared for recklessness than ever.

At the city and state levels, however, movements are afoot to change how politicians raise money and to increase the role of public (that is, taxpayer) funding in campaigns, in place of private donors. Cities like New York and Portland, Oregon, have passed initiatives to dilute the influence of special interests in election bids, and so have states including South Dakota and Missouri. These reforms could help better align the near-term choices that politicians face with the long-run collective good.

In the meantime, well-meaning leaders like Kit Smith need some tricks for planning ahead.

In Homer's epic poem *The Odyssey*, the hero Odysseus, the king of Ithaca, is called away to fight in the Trojan War, and then faces an arduous journey to return home. His travels back to Ithaca to reclaim the throne turn into a series of misadventures involving blinding a cyclops, avoiding the alluring song of the Sirens, and being held hostage by the amorous nymph Calypso. During the two decades that Odysseus is away, his wife, Penelope, withstands overwhelming social pressure. Her husband is presumed dead. More than one hundred men want to marry her, and as time passes, she has a hard time resisting their overtures.

When Odysseus needs to overcome his temptation to succumb to the perilous song of the Sirens, he asks his sailors to tie him to the mast. His tactic is invoked by people today who try to resist their immediate urges to indulge in chocolate cake, for example, by "precommitting" themselves to some painful consequence of giving in to instant gratification, like having to donate money to a hated cause.

But in Homer's story, Penelope has it much harder. It's not her own urges that are the problem—although waiting twenty years for a partner to come home does not sound easy. Penelope is most challenged by the pressure of others. To make waiting for her beloved Odysseus possible and socially permissible, she constructs a ruse to put off impatient suitors. She pledges to marry only after she has woven a funeral shroud for her father-in-law, Laertes. And she makes sure this public marker of her eligibility is never reached. Every day, she weaves some of the shroud, and every night she unravels what she wove during the day. (Unfortunately, a tattletale of a maid discovers her handiwork and exposes her scheme, subjecting her to even greater pressure to remarry.)

The insight to take from Penelope's side of the story is that devices can make waiting possible amid the pressure of those with influence over us. Solon, who was elected the chief magistrate of Athens in the late sixth century BCE, had his own version of such a device. He persuaded the Athenian council to commit in advance to maintain his reforms—to free enslaved people, clean up the local currency, and establish civil rights—for a decade after he enacted them. He also vacated the city-state soon afterward, in voluntary exile, rather than amassing more power, which was when he traveled to see the unlucky King Croesus. He wanted the Athenians to keep the laws intact, and they could not repeal them without his approval. He did not want to hear their Siren songs.

Devices to prevent reckless decisions need not be dramatic; they can even be metaphorical. Harvard professor Graham Allison, who served as President Clinton's assistant secretary of defense, talks about how American diplomats avoided a nuclear conflict with the Soviet Union during the Cold War. U.S. leaders envisioned their standoff with the Soviet Union as "cold rather than hot," he says, a sensory description that led them to choose restraint over rashness, made formal in arms control measures and patient investments to undermine Communist ideology over time.

The irony of the climate crisis today is that it is creating great urgency for communities and societies to act, when we also need to exercise foresight about those actions. Brown University scholars Amanda Lynch and Siri Veland criticize the designation of our epoch as the Anthropocene, an age defined by geologists as one in which humans are dramatically reshaping the planet and pushing its life-support systems to the brink of catastrophe. This now popular name and approach to our time period, they argue, will encourage countries and communities to entertain what may be shortsighted decisions, such as pumping sulfates into the stratosphere or dumping iron into the oceans to cool the planet. They argue for deliberative and inclusive decision making that brings together diverse communities and cultures, rather than isolated reactions.

Whether it's for seeking profit or saving the planet, collective decisions are more and more often being made under pressure-cooker conditions. Like the doctors prescribing antibiotics while fatigued and under time constraints, political leaders and governing bodies are primed for recklessness. They need ways to release the pressure.

As Kit Smith's doubts about Green Diamond began to grow, the investors demanded a quick decision, for good reason. In their business, time is money. Real estate developers rarely want to sit on an empty piece of land, and it gets difficult to rally co-investors if permits and political support lag. Several of Smith's fellow council members, meanwhile, had become enthusiastic about the project, and they did not want to spook the developers lest they take their plans to another community.

Smith, however, decided she'd buy herself and the county time to learn more. She drafted a resolution that offered tentative support for the project, making it contingent on getting more information about the levees as well as the financial and public safety risks. The Richland County Council, following her lead, passed it unanimously. The investors forged ahead and bought the property, seeing it as a green light.

Kit Smith's measure to buy time turned out to be a critical—albeit accidental—device that allowed her and Richland County to resist a reckless decision to develop the land along the Congaree River.

As Smith and the Richland County staff took time to do more research, a piece of local history surfaced. The cost of vouching for levees along the Congaree was not hypothetical. Decades before, the City of Columbia had been sued by the family that owned the same farm, after a 1976 flood breached a levee that the city owned at a nearby site. The city had previously agreed to maintain the levee. The landowner, Burwell Manning, won the lawsuit, which cost the taxpayers of Columbia $4 million.

Weston Adams heard about Green Diamond and knew right away that he didn't like the idea. A thirty-four-year-old Republican whose family had lived in the Congaree River basin going back nine generations, he and his brother grew up hunting for ducks, mourning doves, and wild turkeys in the swamps not far from where the development was being proposed. Adams still owned part of his family's former plantation on the river, which he and his wife later turned into a wedding venue.

Adams plays the part of a southern gentleman perfectly. He has a long, jocular stride and a permanent twinkle in his eye. He looks like he was probably a contender for high school prom king. We met years after the Green Diamond controversy in the basement dining room of the Palmetto Club, an exclusive preserve of Columbia's conservative elite, where white-haired, white-skinned men in dark blazers glad-handed one another, and a lone woman in fur approached to ask if Adams's freezer was full after hunting season. The servers, all black, called the club members by their last names—"Mr. Adams"—but were introduced to me by their first names.

As a lawyer, Weston Adams represented real-estate developers for a living. But to him and his brother Robert, a well-known Republican campaign manager and fundraiser, Green Diamond was an affront to the natural heritage of South Carolina. Weston got a dreamy look in his eyes as he compared the bottomlands along the Congaree to the haunts of Huck Finn, and recalled the adventures of his youth in the swamp. He felt that the land along the river should remain farmland or timberland, not home to a business park with a golf course and shops—and certainly not one that might take its style points from Myrtle Beach. His nostalgia for the past drove his concern for the future.

While Kit Smith and county officials were gathering more data, the Adams brothers enlisted the financial support of Harriott Hampton Faucette, the philanthropic daughter of a newspaper editor who had led a campaign in the 1950s that eventually led to the creation of Congaree National Park. Together with another Republican public relations guru, the brothers joined the fight against the development. They held press conferences every few weeks to counter the developers' media campaign, and they sent out mailers to residents to match the brochures sent by Columbia Venture. Taglines such as "It's a Floodplain, Stupid," and "Your Green, Their Diamond" became rallying cries for a growing opposition to the project that brought environmentalists in the university town of Columbia together with Republican hunters and wealthy landowners.

Smith, a Democrat who at the outset had fought a lonely battle against the project, felt the Adams brothers and their coalition had been sent from heaven. She finally had vocal constituents who supported caution. The public relations blitz that ensued helped her recruit enough support on the council to keep the 1994 county storm-water ordinance prohibiting new floodway development strong, without an exception for Green Diamond. The battle persisted for years because of appeals Columbia Venture made to FEMA to change its flood map in Richland County.

After five years of trying, Columbia Venture failed to get a satisfactory map or waiver to build that would allow it to create Green Diamond on the grandiose scale it envisioned. And so, in 2004, the company sued Richland County for $42 million. Its lawyers argued that the county had deprived the company of the economic value of their land—a violation of the Takings Clause of the Fifth Amendment.

In the late 1960s, the company that owned Grand Central Terminal in New York City had dreams of grandeur. The eight-story railway terminal in the French Beaux-Arts style, which opened in 1913, was already a marvel. Built mostly of granite, Grand Central stretches elegantly along Forty-second Street, its façade featuring a thirteen-foot clock in Tiffany glass flanked by sculptures of Minerva and Hercules; inside, the main concourse ceiling is covered by a night sky of stars. Its Tennessee marble floors and wide stone archways evoke the feeling of walking through an Old World palace. A horizontal edifice of awe-inspiring but not neck-breaking height, Grand Central today sprawls defiantly amid the skyscraper canyons of midtown Manhattan.

For the company that owned Grand Central in the '60s, however, the unoccupied airspace above the station had dollar signs skywritten on it. In 1968, the company entered into a lease with a builder to erect an office tower of more than fifty stories atop the terminal. In one blueprint, the southern façade facing Park Avenue South would be torn down, as would parts of the historic terminal itself.

New York City had just lost another architectural treasure, the original Pennsylvania Station on Seventh Avenue. Its owner had seen a similar opportunity to sell the air rights above the station, and so its demolition began in 1963. The original station, also a Beaux-Arts building, with pink granite walls, a dramatic exterior colonnade, and a main hall modeled after the Roman Baths of Caracalla, would be replaced by Madison Square Garden and a sleek office tower. The station's rail lines and ticket counters would be relegated to an underground maze of confusing low-ceilinged corridors. Of the transformation, former Yale professor Vincent Scully wrote, "One entered the city like a god. . . . One scuttles in now like a rat."

After the demolition of Penn Station, activists in New York advocated for and passed the city's first law to protect historic landmarks, and Mayor Robert Wagner created its Landmarks Preservation Commission. For the company with eyes on reconfiguring Grand Central, however, this presented a hitch. Before the company conjured its plans for the terminal, the Landmarks Preservation Commission had designated Grand Central a historic landmark, and the city blocks on which it sits a landmark site. City law required permission before any property owner could substantially alter a historic landmark.

The commission held four days of hearings with more than eighty witnesses to consider the proposal for a skyscraper to be built above Grand Central. In the end, the commission decided that the plans would dwarf the historic landmark, reducing it to an afterthought. The company sued the city for lost economic value—it had projected earnings of $1 million a year in rent during the construction and $3 million a year in the office space thereafter. The cost to the city to compensate the company would have been staggering.

The U.S. Supreme Court ultimately heard the case, in which lawyers for the company invoked the Takings Clause to argue that the historic landmark law merited compensation to the property owner. In 1978, in a six-to-three decision, the Court decided in favor of New York City and set forth a new test for deciding whether a local regulation was actually a taking. The Court found that takings cases should consider whether the property owner could have had "reasonable expectations" to earn what it claimed. It mattered whether there were other potential commercial uses of the property and if the government policy was protecting the public interest. In his opinion for the Court, Justice William Brennan wrote about the importance of communities protecting the greater public good, for people now and in the future:

Not only do these buildings and their workmanship represent the lessons of the past and embody precious features of our heritage, they serve as examples of quality for today. Historic conservation is but one aspect of the much larger problem, basically an environmental one, of enhancing—or perhaps developing for the first time—the quality of life for people.

The lawyers defending Richland County in the multimillion-dollar law-suit over Green Diamond made the case that the county's storm-water ordinance prevented the public from harm and served the greater good of the community. They argued that it did not deprive Columbia Venture of making other profitable use of the land, through farming or ecotourism, for example. At trial, they aimed to establish that the company had no reasonable expectation to build Green Diamond. Before the company bought the land, the federal map had shown clearly that 70 percent of the property would be in a floodway and the storm-water ordinance had already been on the books for years. They used the Supreme Court's decision about Grand Central Terminal to make their case.

An internal document from the Burroughs & Chapin board, released during the trial, revealed that the company's executives had attempted to reckon with where they had gone wrong with Green Diamond. Among the lessons they recorded were not to pursue a major project outside their home of Horry County where there were "questionable environmental issues" and not to pursue a project "purely speculative in nature."

The South Carolina judge presiding in the trial over Green Diamond ruled in favor of Richland County. The developer, Columbia Venture, appealed to the State Supreme Court, which took up the case. In August 2015, the court affirmed the lower court's decision: Richland County did not owe the developers any compensation for its decisions or for its laws. The justices saw these actions as protecting the future interests of the public.

When Kit Smith was on the witness stand, the cross-examining attorney for Columbia Venture suggested that she had played political games in maneuvering to prevent a world-class development. Smith disagreed. In her view, she had simply bought time to deliberate, and to learn about the future dangers of developing in that floodplain. With more time, she found a way to bring the rest of the council around to her view and to marshal political support from a coalition of allies that crossed party lines.

Weston Adams, the Republican hunter and lawyer, and his friends, when allied with liberal environmentalists, made for a potent force resisting Green Diamond. When people are willing to align with strange bedfellows, they stand a better chance of outmatching short-term special interests. Within smaller communities, where people can meet face-to-face and earn trust despite their differences, this can be easier than in national politics. The intimacy of South Carolina politics had two faces—more intense social pressure for its leaders, but also more opportunity to build coalitions.

Another trick to overcoming community myopia was having leadership both fierce and financially independent enough to stand up to the interests of short-term speculators, and also in having devices to delay instead of make reflexive decisions. To exercise foresight was to endure criticism nonetheless. The legal victory for Richland County showed that courts can deter reckless decisions. But courts can also incite recklessness, directly or indirectly, as with the trend that followed the 1922 Pennsylvania Coal decision.

Every community and society has such choices. Laws and institutions can actively discourage myopia or encourage it, at every level of government. The Gulf Coast red snapper fishery described in chapter 5 offers another example of how government policy can either constrain looking at the road ahead, as the derby days did, or aid it, as the catch share scheme did. Valuing the future demands that we rethink our rules and preserve institutional integrity. These collective agreements have the potential to guide our gaze back to the horizon.

The victory for Richland County came just a few months before the rains. In October 2015, the South Carolina midlands experienced nearly a week of torrential downpours, breaking historic rainfall records.

Water breached the walls of the Columbia Canal, contaminating the city's potable water supply. The floods sent caskets floating away from cemeteries. Businesses along Bluff Road, a route that runs parallel to the Congaree River, were submerged in water neck deep. Five dams broke on Gills Creek, the tributary to the Congaree that runs across the former Green Diamond property, inundating the land in muddy water that lingered for days. Thousands of people lacked clean drinking water during that time, many families lost homes and businesses, and several people drowned. The tally of property damages in the state rose to $12 billion, devastation met with federal disaster relief.

Despite the destruction, the rains had not wrought anywhere close to the damage likely to happen when the Congaree River floods and breaches its levees—an event that grows more likely each year on a warming planet.

"When I think about what could have been," Lisa Sharrard, South Carolina's former floodplain coordinator, said, "the hairs stand up on the back of my neck. Chills go through my spine. If they had let Green Diamond happen, the flood could have been a catastrophe. Hundreds or even thousands of people could have lost their homes and their jobs at one time." Sharrard showed me a photograph from October 2015 of a flooded highway on-ramp near the former Green Diamond property, and then we got in her car and drove to the same spot. The water had reached the height of the semi-trucks skimming below the overpass. It had rushed like a mighty river, drowning those in its path.

THE GAMES WE PLAY

Rehearsing the Past and Future

Time is a great teacher, but unfortunately
it kills all its pupils.

—HECTOR BERLIOZ

When Georg Sieber woke in Munich in the summer of 1972 to his phone ringing at a quarter past five in the morning, he immediately suspected what had happened. When he picked up the phone, his fears were confirmed.

He quickly dressed and jumped on his moped, speeding across town to where team officials and police had gathered outside a makeshift village that temporarily housed twelve thousand athletes from around the world. There he learned that a group of extremists clad in tracksuits had jumped the fence of the Olympic Village and taken nearly a dozen Israeli athletes and coaches hostage. For Sieber, this news did not come as a surprise.

The thirty-nine-year-old psychologist for the Munich police had light eyes punctuating inquisitive eyebrows, a rugged chin, and an elegant manner of gesturing as he puffed a cigarette. He earned his post by first serving as a double agent, infiltrating student protest groups in Munich and informing the police of their planned demonstrations. He saw no moral contradiction in marching in the 1960s with the radical Students for a Democratic Society to demand better housing conditions while also quietly divulging his comrades' whereabouts to the police. In fact, he believed his work fostered the predictable circumstances that would prevent violent confrontations between police and protesters. If law enforcement officers could anticipate where the demonstrators would be, Sieber thought, they might not react rashly by spraying tear gas or shooting at them.

In the months before Munich hosted the 1972 Olympic Games, Sieber was charged with the task of imagining scenarios of what could go wrong. Since he knew the city and anticipated crisis situations for a living, he became an adviser to the Olympic security team. The twenty-six scenarios he concocted ranged from a spectator pushing another spectator into a swimming pool to an attack by a Swedish terrorist group flying a plane into the Olympic stadium. He considered possible plots by the Basque separatist group ETA and the Irish Republican Army. (The Olympics, after all, celebrated the world's recognized nation-states and not their subnational groups.)

In Munich, Germany would host the Summer Games for the first time since Hitler had presided over the Berlin Olympics in 1936. The hope of the organizers was to defy the stereotype of Germans as morose and militaristic, and rebrand the country as carefree and Mediterranean. The Germans called them "Die Heiteren Spiele," which roughly translates as "the cheerful games." The mascot was a multicolored dachshund named Waldi.

What became known as Situation 21 did not arise solely from Sieber's imagination. He drew on news reports he had read about Palestinian extremist tactics. (The terrorist group Black September had carried out several operations in Europe in the previous year.) He also contemplated the dynamic of the Games themselves. Only recognized nations could send athletes; the requests of Palestinian groups to send an Olympic team to Munich had been in turns ignored and rejected.

With uncanny prescience, Sieber imagined the extremists' taking Israeli hostages. In his report to the organizers, he envisioned that terrorists would climb the unsecured fence of the village before dawn, corral hostages in the complex, and make demands. The terrorists, he projected, would not surrender and would be willing to die to make their political statement. Sieber anticipated this scenario before Black September had even planned its actual attack.

Simple precautions might have prevented Situation 21 from unfolding in reality. The Olympic organizers failed to take even the most basic and least expensive security measures, such as putting barbed wire on the village's perimeter fences and arming police or troops to stand guard. The athletes might have been housed by sport, rather than nationality, as suggested by Sieber, to make it more difficult to target Israelis.

Instead, the organizers of the Munich Games asked Sieber to dial down his terrifying scenarios so they might be less apocalyptic, and more befitting the carefree games they intended to host.

A powerful factor was at play in the Olympic organizers' dismissal of Sieber's scenarios: history. The collective memory of the horrors of the Holocaust and Hitler's reign over the 1936 Olympics was still present and visceral for people all over the world, particularly in Germany. Many Germans who had survived the Third Reich felt with salience the shame of having been complicit in one of the most destructive and repugnant regimes in history. There was an urgent desire to compensate for the past, to make the Games feel lighthearted and to combat the perception of the country as a police state.

We still see traumatic memories shape decisions in our era, including in the United States. Before 9/11, even insurance companies, whose business is calculating future risk, neglected to account for the extraordinary financial risks of a terrorist attack in their policies to insure buildings in New York. This was despite the World Trade Center bombing in 1993. Wharton economist Bob Meyer's research shows that after 9/11 many companies began paying exorbitant premiums to protect against terrorist attacks, beyond what was warranted. The horrors of that day in history were seared not just into people's memories, but into their projections of the future.

Just as memory can warp our view of the future, the lack of memory—or collective amnesia—can drive reckless decisions.

Six years after the devastating nuclear disaster in Fukushima, Japan, I traveled to the region with a small delegation of leaders from the United States. We met with executives from the Tokyo Electric Power Company (TEPCO), which ran the infamous power plant known as Fukushima Daiichi. The company was still working to clean up and decommission the station after the nuclear reactor explosions, a task likely to endure for decades. I was astonished to learn that the company was trapped by its views of the past, in more than one way.

On March 11, 2011, a 9.0 earthquake off the shore of Japan triggered tsunamis that breached the seawall protecting the Fukushima Daiichi nuclear plant, flooding its backup generators and pumps that brought cooling seawater into the reactor. Three reactors heated up until they exploded. It was one of the most destructive nuclear disasters in history, displacing hundreds of thousands of people. For months after, area school children wore dosimeters on their wrists to gauge their radioactive exposure, and doctors examined thousands of children for potential thyroid cancer.

When I visited in March 2017, more than a hundred thousand people had yet to return to their homes in Fukushima; some lived in trailers adjacent to nuclear waste storage sites. Monitoring stations were set up at playgrounds to measure exposure to radioactive material. The economy of the region, long reliant on farming of rice and fruit and known for specialty products such as gourmet apples and sake, had been crippled by people's fears of eating food from the site of the meltdown—even if the cultivation happened in the mountains, at high elevations of the prefecture far from the fallout zone. Japanese authorities shut down dozens of nuclear reactors around the country for safety concerns. Few new ones were being built, leaving the country more reliant on natural gas and other fossil fuels.

To be sure, the earthquake of March 2011 was the most severe in the recorded history of Japan. But earthquakes were expected in the region, and what happened in Fukushima was foreseeable. The tsunami that breached the seawall at Fukushima Daiichi was a one-in-a-thousand-year event, but it had not been included in the simulations that the power company, TEPCO, used when building the plant. Nor were updates made when new information became available. In 2002, TEPCO ran an analysis revealing that the tsunami risks at Fukushima had been underestimated and that its seawater pumps to cool the reactor were in danger.

The company did not make any changes to respond to this realization, nor did its government licenses explicitly require changes as a result. Officials in the Japanese nuclear industry and its overseeing government agency were focused on earthquake risks, neglecting the historic (and related) risk of tsunamis in Japan. Studies of tsunamis dating back to the fifteenth century would have revealed that massive waves could breach the power plant's seawall. A historic earthquake and tsunami in the year 869 in Japan documented by researchers also proved the risk of a massive tsunami in the region. Yet the wall at the power station was not built to withstand waves of even half the height of those that hit Fukushima in March 2011, according to an analysis by the Carnegie Endowment for International Peace.

A nuclear plant in a neighboring area, meanwhile, had been built to withstand the tsunamis. A solitary civil engineer employed by the Tohoku Electric Power Company knew the story of the massive Jogan tsunami of the year 869, because it had flooded the Shinto shrine in his hometown. In the 1960s, the engineer, Yanosuke Hirai, had insisted that the Onagawa Nuclear Power Station be built farther back from the sea and at higher elevation than initially proposed—ultimately nearly fifty feet above sea level. He argued for a seawall to surpass the original plan of thirty-nine feet. He did not live to see what happened in 2011, when forty-foot waves destroyed much of the fishing town of Onagawa, seventy-five miles north of Fukushima. The nuclear power station—the closest one in Japan to the earthquake's epicenter—was left intact. Displaced residents even took refuge in the power plant's gym.

Still reeling from the disaster when we visited, TEPCO officials described to our delegation that they were now going to create the "safest" nuclear power station in the world, a reboot of the Kashiwazaki-Kariwa nuclear plant on the west coast of Japan, in a prefecture that shares a border with Fukushima. The safeguards they described to prevent future disaster included a forty-nine-foot-high seawall, a twenty-thousand-ton water reservoir for cooling the reactor, forty-two fire trucks, and twenty-three generator trucks as backup for making sure a reactor can be cooled in an emergency. Emergency drills would be held every month. It was clear that they were preparing the upgraded station for the exact disaster that they had recently experienced in Fukushima and were still cleaning up.

A former senior Pentagon official and expert in energy security from the U.S. delegation asked the TEPCO executives if they were preparing for risks to the nuclear power plant other than a tsunami. The power company officials returned blank stares. She pressed them about potential attacks from North Korea, which earlier that week had launched a missile into Japan's exclusive economic zone. They said they had not considered it. Nor had they evaluated scenarios of sea level rise that might raise the level of storm surge from future cyclones, or heat waves due to climate change. In other words, the "safest" nuclear power plant would really be safe from the specific horrors of the recent past. A Japanese government official who later heard about this exchange shared the U.S. official's alarm.

What remains ablaze in memory guides how we prepare for the future. What fades from memory, by contrast—even in places that have experienced horrifying disasters—can also fade from future imagination.

When Seneca wrote *Natural Questions* immediately after the earthquake in Pompeii in the year 62, he described residents living in fear, many of whom had fled or were deciding whether to flee the region.

Seventeen years later, according to Pliny the Younger's account, fears of the shaking earth had receded from memory, and thousands ignored the tremors before the volcanic eruption.

People similarly refuse vaccines at higher rates at moments in history when there are fewer cases of the targeted disease. Popular movements to refuse smallpox vaccination gained momentum after a period of low prevalence of smallpox in London in the 1850s, and similarly in the United States in the 1830s. This is because the devastation wrought by disease fades from a society's memory with more widespread inoculation, argue epidemiologists Saad Omer and Diane Saint-Victor, who have looked at the history of global vaccine campaigns dating back to nineteenth-century England. Omer and Saint-Victor believe that as a result, it's the "last mile" of vaccination, reaching the full scope of the population amid rising refusal and dropping disease rates, that is the most difficult.

Immediately following a disaster, there is a surge in people buying hazard insurance. For example, after the 1994 earthquake in Northridge, California, two-thirds of nearby residents had insurance by the next year. No severe earthquakes have hit California since then (as of this writing), and Wharton economists Bob Meyer and Howard Kunreuther have shown that recently, less than 10 percent of homeowners in the high-risk areas of the state have earthquake insurance. Residents will likely be unprotected when the big one hits, with damages that could cost hundreds of billions of dollars.

We pass along our collective amnesia to predictive models that are meant to help us calculate future risk. Even risk experts make the mistake of constraining computer simulations to the time period for which they have robust data, or they don't look back far enough across history. The tsunami simulations used by TEPCO before building the Fukushima Daiichi Nuclear Power Station did not account for the span of historic data on massive waves in the region, just the earthquakes and tsunamis in most recent memory.

In the lead-up to the 2008 financial crisis and Great Recession, Moody's, one of the three major agencies that gave mortgage-backed securities a top-notch AAA rating—assuring their high degree of safety as investments—used models with American housing data going back only twenty years when estimating the risk of default. Nate Silver has pointed out that the period, when home prices rose and overall mortgage delinquencies were low, did not convey the true risk because it did not show how mortgage defaults could be correlated with one another, causing chain reactions. The agency neglected information from decades prior, including the Great Depression, which could have shown the danger of people defaulting on mortgages as housing prices plummet. In 2007, delinquencies spread like contagion through the housing market, and the high-risk securities crashed, crippling the economy. Two ratings agencies, Moody's and S&P, had rated the securities as having a risk of default two hundred times less than what they actually had. Economists and policy makers virtually ignored the risk of a recession. (The agencies, of course, also had perverse financial incentives, with investment banks paying them fees to rate their financial instruments, including high-risk mortgage-backed securities and collateralized debt obligations.)

History clearly deceives us about the future. But can it also do the opposite—prepare us for threats that lie ahead? The Athenian general Thucydides wrote in his oft-cited history of the Peloponnesian War between Sparta and Athens that "the present, while never repeating the past exactly, must inevitably resemble it. Hence, so must the future." Mark Twain is rumored to have said that while history does not repeat itself, it rhymes. A proverb in Madagascar, meanwhile, advocates that we act like the chameleon, with "one eye facing backward at history, the other facing toward the future." There's wide appeal, across cultures and epochs, to the idea of shedding collective amnesia, and the idea that it might actually be good for us to remember the past.

But in what way does the future resemble the past, and how should we use history to avert catastrophe? History did not help the French military in the 1930s prepare for World War II. The French assumed the Germans would fight them in trenches along the Western front, as they had during World War I. They left the northern Ardennes forest unguarded while fortifying the Maginot Line, and France fell to the Nazis within the span of two months.

In the fall of 2016, before he became the U.S. secretary of defense, retired four-star Marine Corps general Jim Mattis agreed to speak with me. I sought him out because of his reputation for thinking across long spans of time. History, in his view, should be used to stretch leaders' perspective of what is possible. Mattis had a library of more than six thousand books and exhaustive—at times, exhausting—knowledge of military history. He lambasts Marines who complain about lacking time to read, and praises a Corps that now prescribes more history books as people rise up the chain of command from captain to major to general. He credits a historical perspective with his own success. "How do you get to four stars? You fight enemy generals who are dumber than a box of rocks," he quipped while driving among the ancient redwood trees of Northern California.

I found Mattis compelling even though I don't share his politics or his penchant for fighting. He has been called the "Warrior Monk," which seems an apt way to capture his wisdom and ferocity, both apparent when we spoke.

War is not in any way predictable, Mattis acknowledged. "So then why should warriors read history?" I asked. The particulars of how a war unfolds are chaotic, and not something that people should try to forecast in advance, he said. Weapons change over decades and centuries. But he believes the drivers of war are immutable—the same as they have been since the Trojan War. "Nations go to war over fear, over honor, and over interest," he said, paraphrasing his own favorite line from Thucydides. Keeping history alive in collective memory, he believes, could help countries defend their interests more wisely.

Mattis adamantly and vocally opposed the 2003 U.S.-led invasion of

Iraq as a response to the terrorist attacks of 9/11, in part on the grounds of historical perspective. To succeed in fighting a war there, he believed the United States' political leaders needed to be able to imagine a realistic end state in Iraq after the invasion. In his studies of the region's history, he saw no basis for imagining an inclusive democracy among warring factions in Iraq. It was an ahistorical hallucination to think that Sunni and Shia Muslims with long-standing tensions would unite in a post-invasion Iraq.

The lack of an endgame for Iraq, Mattis believed, should have been a red flag given the U.S. experience in the Vietnam and Korean wars, where it also lacked a shared vision among military and political leaders of an end state. He contrasts this with the 1991 Gulf invasion conducted by President George H. W. Bush, who defined in advance an end state—to push the Iraqis out of Kuwait and go no further. In 2003, Mattis said, "Fear drove us. We went helter-skelter and we didn't do intelligence planning. We didn't even make sure the wounded could be evacuated."

Mattis did not prevail when presenting the historical view on the proposed Iraq invasion to the political leaders of the latter Bush administration. But he still followed orders. His field officers during the invasion were required to read *The Siege of Kut-al-Amara*, about the defeat of the British-Indian garrison south of Baghdad by Ottomans in World War I. The siege endured more than five months. The history of that long-fought battle could not predict or prescribe what the officers would face in Iraq, but Mattis believed it gave them a way of understanding blurred lines between enemy and ally in the history of the Middle East.

Former Harvard professors and government advisers Ernest May and Richard Neustadt tried to bring greater rigor to the idea of using history to make wise decisions about the future. It's best to use multiple historical analogies when making a decision, they argue, not rely on single past precedents. They point to how President Harry Truman and his secretary of state, Dean Acheson, made the decision to intervene in the war in Korea in 1950, drawing on the history of not just one country or situation, but incidents from Ethiopia, Manchuria, Greece, Berlin, and Austria across centuries. They draw a sharp contrast between that use of many historical analogies, each scrutinized for their relevance, to the way Lyndon Johnson relied merely on France's invasion of Vietnam in 1954 as a historical analogy for the U.S. war in Vietnam.

In their seminal book *Thinking in Time*, Neustadt and May laud President John F. Kennedy's decision making process during the Cuban Missile Crisis of 1962, a confrontation between the two Cold War superpowers, as a model for how political leaders should use history. Thirteen days of tense deliberation focused on how the United States should respond to its discovery that the Soviets were transporting nuclear missiles to Cuba, just ninety miles from the United States.

The ExComm, as Kennedy's advisory group was known, by the end of its first day of deliberations began to actively imagine how its decision would be viewed over the course of history. (This, May and Neustadt tell us, is a very rare practice in most political decision making; my experiences working in government corroborate this.) By making an air strike on a Soviet target—an option advocated by some officials—the United States might have started World War III, and possibly even triggered a civilization-ending nuclear apocalypse. Members of the ExComm paired active imagination of how they would be viewed in history with lively debate about whether particular historical analogies—such as Pearl Harbor—applied to the current predicament. The committee was composed of experts of differing ideologies—some hawks and some doves—who had deep historical knowledge specific to both Cuba and Russia, and who could think through the cultural and political dynamics at play. President Kennedy had also learned from his mistakes during the 1961 Bay of Pigs invasion; he recused himself from conversations to create more opportunity for open disagreement during the missile crisis.

The lesson, it seems, is to look at many historical events as possible precedents for the future. Leaders of communities and societies also might do well to bring together diverse people who openly disagree, to strive to keep any single history from dominating the collective view of the future.

When the Jogan earthquake struck Japan in the year 869, residents of Miyako-jima fled to the top of a hill on their island. They did not flee to safety. A giant tsunami rose over the hill from one direction, while a second massive wave surged from a rice paddy on another side. The two waves collided, crashing on the hilltop, and swept the people seeking refuge at its pinnacle out to sea. The waves destroyed the fishing village of Murohama.

The horrific tragedy was not soon forgotten. Its memory lingered for more than a thousand years. A stone marker on the hill, placed next to a shrine, told the history of that day and warned future generations not to seek shelter in that place. The warning entered local folklore and was studied by children in local schools. When the earth shook in March 2011 in Japan, nearly all of Murohama's inhabitants remembered what had happened 1,142 years before. They heeded the warning and fled farther inland, and watched two giant waves crash over the hilltop again.

In Aneyoshi, Japan, an aging stone tablet stands sentinel, warning future inhabitants, "Do not build your homes below this point!" No homes stood below its elevation in 2011. The waves of the recent tsunami lapped just three hundred feet below the marker.

These two communities are exceptions, not the rule. Hundreds of other stone markers commemorating tsunamis are scattered across Japan, many erected after devastating tsunamis in 1896 and 1933. According to a study by the Nuclear Energy Agency of the international Organisation for Economic Co-operation and Development, virtually none of the other communities with such markers heeded them as Aneyoshi and Murohama did.

What made these villages, but not the others, heed history?

The communities in Japan where the historic markers made for effective warnings were small villages, with cultural continuity across generations. Schoolchildren learned the history of the past tsunamis and the need for vigilance. And the stone markers in Aneyoshi and Murohama stood out relative to the hundreds of others in Japan for offering specific actions rather than just vague commemorations of history: Do not build homes below this point. Do not flee to this hill.

When profound past experiences provide precedent for a credible future scenario, history can serve the present. Drawing on historic incidents can revive a memory and keep a threat salient, as it did for the Japanese engineer who knew the story of the ancient tsunami that had flooded his hometown's shrine. In 2016, I spent some time in an American community that is attempting to rescue the past in this way for the purpose of keeping a threat alive in memory. But the community is not simply building memorials to past disasters.

Mattapoisett is an intimate coastal town of six thousand people in southeastern Massachusetts just east of New Bedford. Its eponymous river empties into a pocket of Buzzards Bay. The town's main drag, populated by historic homes and a park with a grandstand, runs parallel to the seashore. A deep harbor and abundant timber, as well as proximity to the whaling industry's epicenter of New Bedford, attracted shipbuilders in the 1750s from Boston's south shore who settled and built the village. The town hall is perched less than five hundred feet from the water.

In the twentieth century, two historic hurricanes pummeled Mattapoisett. The 1938 hurricane predated the naming of tropical cyclones in the United States but distinguished itself nonetheless. The sea swept away houses, flooded the streets, and washed boats ashore. Century-old elm trees fell across the roads like matchsticks. Floodwaters surged above the steps of the town hall and flowed across the first floor, leaving an indelible high-water mark on the town safe. A thirty-foot wagon shed blew into a field. Henhouses exploded. A sixty-foot yacht washed ashore, and a mansion on the beach was flattened. People were found pinned alive under collapsed roofs and walls of buildings. The local newspaper, in the aftermath, said the storm had turned the summer beach colonies at Crescent Beach and Pico Beach into wastelands. On top of one of the shredded cottages, relief workers found a phonograph record, "Smiles." Hundreds of homes were destroyed, and hundreds of people were injured. Several people drowned.

In 1991, Hurricane Bob, a category 2 storm, howled through Mattapoisett. Waves more than twelve feet tall crashed onto its shore. The storm demolished wharves, lifted beach cottages, and felled trees. Winds sheared and toppled telephone poles. A water spout erupted on the town's golf course, and streets became fast-moving rivers with water rising. People waded through floodwaters to leave the homes they had failed to evacuate, cutting themselves on shattered window glass. The town had no sewer system then; street water became toxic with sewage and the drinking water contaminated with household chemicals. On Cove Street, which traverses a narrow peninsula that juts into the bay, twenty-nine of thirty-seven homes were destroyed.

Even in a town that relishes its storied past, however, the memory of these hurricanes has recently been fading. "The concern I have is that so much time has passed that many people do not fully comprehend the danger," Mike Gagne, the town administrator, told me in 2016. In recent years, when faced with storm warnings and evacuation orders, many residents have not taken precautions. Gagne believes it is only a matter of time before another big storm hits Mattapoisett. He fears that townspeople won't be prepared, whether by elevating their homes in advance or by evacuating and securing their valuables when they see forecasts of an incoming storm.

Officials from the U.S. Environmental Protection Agency at the time of my visit were also worried about the lack of preparation in coastal communities, especially given the growing risk of catastrophic tidal surge due to rising sea levels. Mattapoisett has been a particular area of concern in the New England region because its drinking water supply can be compromised by storm surges. Along the Atlantic Ocean from the Carolinas to New England, the sea level is rising three to four times faster than the global average. If and when a category 3 or category 4 hurricane hits Mattapoisett, it will likely do far worse damage than Hurricane Bob.

Jodi Bauer, a resident whose great-grandparents settled in Mattapoisett in 1908, had an idea for how the town might revive its memories of the past. Bauer is the town barber, the third to hold the job since the barbershop opened in 1928 and the first woman. Her mother, born during the 1938 hurricane, worked as the town clerk for thirty-three years. Her father was a police officer. Jodi spent her childhood at the town hall, and remembered hearing stories about that hurricane. Later, as an adult, she experienced Hurricane Bob. "It was like *The Wizard of Oz*," she told me as we drove around town together on a frothy October morning, the air smelling like damp beach towels and the gardens lush with late blooms. "Some houses got smashed, and some got lifted into the air."

Bauer's idea was to mark the height of the floodwaters from the two historic hurricanes in prominent places around town. A mother of two Boy Scouts, she shared this idea with a local seventeen-year-old high school senior, Jared Watson, who decided to take it on as his capstone project as an Eagle Scout.

Tall, studious, and captivated by science, Jared was the perfect person for the job. He worked with the city to create metal bands and striking blue signs to be drilled into electric utility poles at major intersections and landmarks around town. The bands marked the height that floodwaters reached during the 1938 and 1991 storms; the signs provided a key to interpreting the bands and pointed passersby to scan a bar code to learn about the growing risks of coastal flooding with sea level rise. Scientists from the EPA collaborated with Watson and the town to calculate historic flood heights and to make future projections.

"These markers convey the history of storms in a more tangible form," Jared told me on the day that he had recruited fellow scouts and community volunteers to help him install the flood markers. I saw that the markers brought the history of the hurricanes into the present, inviting them to imagine the water rising—not in some distant place, but in the familiar places they traveled every day. Gagne says the signs seem to be having a greater impact than handouts and public announcements that he and other local officials have made, because people see them when they drive around town—they serve as more frequent and poignant reminders.

Jodi Bauer had also been collecting oral histories of the 1938 hurricane from town elders, many of whom experienced the storm as children and remembered it vividly. Videos of their stories were shared in the local library and online, and in community events the town leaders have brought these stories to larger groups of residents. It remains to be seen how much the residents of Mattapoisett will prepare for the next big hurricane after this experiment. As I joined crews installing the markers around town, I noticed they sparked many questions from residents driving or walking by. They had started conversations and, in some cases, debates, about how high the waters rose during historic hurricanes.

Like the Japanese villages that heeded the tsunami warnings from the past, Mattapoisett is of a size and intimacy that seem to give it an edge in using history to prepare for its future. The town had also found a way to bring generations together—the young scouts invested in their own future, and the elderly storm survivors who remembered the past—in common cause.

What can we do when we need to imagine a future we have never experienced in our own lives—or in history? Risk expert Nassim Taleb argues that it is the black swans—the unprecedented and unpredictable events—that lead to the most disruptive change. "Living on our planet, today, requires a lot more imagination than we have," he writes. Or, as Stanford scientists Katharine Mach and Miyuki Hino put it after Hurricane Harvey devastated Houston in the summer of 2017 with a days-long deluge, "Unprecedented is increasingly the norm."

On September 5, 1972, two Israeli athletes were killed and nine more were captured at the Olympic Village in Munich by the Palestinian terrorist group Black September, which demanded that 234 Palestinians be released from Israeli prisons. By the next day, all nine hostages had been killed by their captors.

What happened was unprecedented for the people involved—there was not some obvious historic precedent, at least not for the Olympic organizers and the Munich officials, despite the terrorist attacks elsewhere in Europe.

When Georg Sieber arrived on the scene that morning to where the authorities had gathered, he encountered chaos. Personalities from various law enforcement groups clashed, and the lines of authority were unclear. He grew frustrated that his advice on how to respond to the crisis did not prevail. He quit his post within a few hours. Several hours later, the hostage-takers massacred the Israeli athletes. The cheerful games ended in a shootout on an airport tarmac televised for the world to witness.

Sieber did not predict precisely the future, nor did he see it with clairvoyance like the mythical Cassandra, whose curse the ancient Greek playwright Aeschylus described as both foreseeing the ruin of Troy and having everyone ignore her prophecy. Sieber foresaw only potential futures.

Months earlier, when looking at Sieber's original scenarios, the organizers in Germany clearly had not grasped what it would be like to experience the horror of Jewish athletes being kidnapped and killed on German soil at the Games. If they had, some basic security measures would have been seen as well worth the hassle.

I t's not possible to plan for every hypothetical scenario of the future, especially when the costs of doing so are high. Most risks that lie in the future are not like nearby hurricanes that can be charted by credible forecasts, but are rather, by their very nature, unpredictable—the shenanigans of what scientists call dynamic systems. That someone imagined a future that eventually played out does not mean anyone actually knew for sure it would happen.

Ostensibly, the 1972 Olympic organizers in West Germany acknowledged the need to take into account possible future risks. They had tried to elicit imagination of scenarios by engaging an expert. That might seem, from what we know about imagination, like a promising way to avoid recklessness.

But the effort failed. Sieber gave the organizers threats to plan for, and portrayed them in colorful detail. The security officers and organizers of the Olympics refused to even consider certain scenarios.

Why did Sieber's scenarios fail to persuade anyone to take precautions? Why, for that matter, do countless people like him—and me—fail to persuade people to see the pictures of the future we paint?

Scenarios trigger people's mental immune systems to fight off the futures they either do not want to experience or do not believe they will experience, says Peter Schwartz, the founder of the Global Business Network, a think tank that consulted corporations. Schwartz, in decades of scenario planning, has discovered that people are tempted to try to lock in on a single possible scenario that they prefer or see as most likely and simply plan for that—defeating the purpose of scenario generation.

The philosopher Peter Railton, who studies moral cognition and prospection, told me that trying to think about the future is difficult because there are so many possibilities—or degrees of freedom, as he likes to put it. In thinking about the near term, like what we could eat for lunch today, our cognition can be uncontrolled and spontaneous. By contrast, it takes a great deal of mental control to think about future events, and the practice becomes harder with greater temporal distance. The more possibilities we see the future having, the harder it is for us to relate to any of the given outcomes and believe in their plausibility. "We're much better at thinking about a specific future event, pulled out of a bin of all possible outcomes," he says. The conundrum this presents is that when we consider a singular scenario, we run a high risk of ignoring other possibilities.

It's fair to assume that the organizers of the Munich Games never felt the scenarios Sieber imagined in a visceral sense. They did not experience them the way I felt myself swimming in a coral reef in virtual reality. The scenarios remained abstractions, easy to dismiss.

Sieber also described several doomsday scenarios. When people are presented with dystopic futures, it can actually lead them to further privilege immediate concerns and discount later consequences. When we feel we lack control over impending doom, we may give up and seize the moment, says Thomas Suddendorf, the Australian evolutionary psychologist who is responsible for much of our current understanding of how humans evolved to have foresight. "When people are primed for a negative future, a lot of people think: 'Well, it's all going to custard, and I can't do anything about it anyway,'" he told me.

The reaction people have to doomsday scenarios might explain why Sieber was asked to paint futures that looked less like nightmares for the "cheerful" games. Or the officials may have just preferred to take no action. After Dr. Smith Dharmasaroja, the head meteorologist of Thailand, advocated in 1998 for creating a network of sirens to warn of coming tsunamis in the Indian Ocean, the ruling government replaced him. His superiors argued that a coastal warning system might deter tourists, as they would see Thailand as unsafe. Six years later, a massive Indian Ocean tsunami killed more than 200,000 people, including thousands in coastal Thailand, many of them tourists.

Scenario planning is highly popular in both government and business. It makes communities and societies feel like they are exercising foresight about the future. But groups of people, especially when they share a culture or common cause, often reinforce each other in dismissing an unprecedented future risk or opportunity. This is the mirror image of Cameroonian parenting or professional poker, where foresight is encouraged by cultural norms. Instead, some cultures discourage heeding future threats, because leaders or groups favor the present for reasons that span politics, profit, and personal preference. Often, they are unaware of alternative cultural practices that could help them overcome their myopia.

When people are forced to play an actual role in an imaginary future scenario, however, Peter Schwartz has observed that it helps penetrate their natural defenses. Schwartz spent a chapter of his career at the oil company Royal Dutch Shell helping executives and managers plan for scenarios such as oil price shocks and glasnost, the opening up of the Soviet Union under Soviet leader Mikhail Gorbachev. Schwartz describes a period in the early 1980s when he had been struggling to persuade Shell's managers to take extreme scenarios, such as a collapse of OPEC, seriously. A breakthrough came when he gave the group various computer-simulated scenarios of the future and asked them to play the roles of key figures in each situation, such as oil ministers of the OPEC countries Iran and Saudi Arabia, and executives of competing oil companies. By the end of the exercise, the group's executives were taking seriously risks they had been too skeptical to consider before—including an invasion of Kuwait by Iraq, an event that eventually happened in 1990. Schwartz called it a rehearsal. It struck me that the routine, unlike the process used by Sieber, resembled a theatrical production or a game more than an effort to plan for the future in reality.

When I first encountered Pablo Suarez, he had just breezed past the Secret Service with a Frisbee and a pair of giant fuzzy dice tucked under his oversize suit jacket. I'd invited him to the White House in 2014, where I was serving as senior adviser for climate change innovation and trying to help communities plan ahead for weather disasters. Suarez had experience working with the global Red Cross team to help communities around the world prepare—and not just react to—droughts, floods, and tropical storms. His goal was to shift humanitarian aid efforts toward disaster prevention instead of just disaster relief. Lanky and outspoken, with a thick Argentinean accent, a graying beard, and a mischievous glint in his eyes, Suarez struck me as someone out of his element in the staid halls of Washington. I could more readily envision meeting him as a hitchhiker on a country road—all of which made me suspect it was worth listening to him.

It's one thing to be told about a future threat, Suarez told me. He was speaking of the predictions that aid workers share with farmers to prepare for a drought, for example, or those scientists share with development banks to persuade them that it's worth investing in planting trees upstream of a river that floods. It's far more profound—even moving—for groups of people to experience in some way the emotional consequences of the decisions they could make about the future. Suarez believes that such experiences can be simulated by simple games. A research fellow at Boston University's Pardee Center and the associate director of the Red Cross Red Crescent Climate Centre, he has designed dozens of games that have been played by more than ten thousand people around the world, including subsistence farmers in sub-Saharan Africa, weather service officials, international humanitarian donors, politicians, and insurance industry executives. The goal of the games is to help people inhabit—indeed, almost experience firsthand—what it is like to act on predictions of future risk.

In Senegal, for example, Suarez and his team played a game with inhabitants of an island where people die often from storms that can be anticipated. The players of the game drew cards that presented a situation—a forecast of an incoming severe storm, for example. Players then chose from action cards, such as sending their kids to their grandparents' house or seeking better shelter. Then they saw what the outcome of the forecast was, and how it played out for them given the decision they made. Repeating the game made clear that a storm does not cause havoc every single time, but that when a severe storm strikes, it has devastating yet often preventable consequences.

In another game he plays with disaster relief workers, Suarez tries to demonstrate what happens when, in light of an ominous forecast, they preemptively distribute relief supplies like tents to a region, versus waiting to see what happens. The players in the game start with a handful of dried beans. Beans can be used to pay for forecasts, or to pay for supply distribution before or after an extreme weather event. The players make a decision about what to do, then roll dice to see whether floods strike in each circumstance.

The cost in beans in the game of taking the precautions is much less than the cost of reacting to a disaster after the fact. This simulates a dynamic in the real world, where the cost of responding to disasters, when bridges may be washed away and roads become impassable, far exceeds investments to prepare in advance.

In his experience with humanitarian aid organizations, Suarez has found that groups fear acting in vain—that is, spending time and money taking precautions based on a scientific forecast when maybe no severe flood or famine will actually strike. Yet people also regret not preparing in advance when disasters do strike. The games acknowledge that it's possible to gauge the eventual consequences of choices before we experience them, but it's difficult for the average person to deeply understand the possible outcomes. His games aim to help people muddle through that uncertainty by seeing and feeling what can happen.

Suarez believes that games help communities to understand and feel some of the consequences of loss from disasters that most of them have not experienced in their lifetime, or at least in recent memory. Within a game, you can simulate how small decisions—not investing in forecasts or insurance, or not planting trees that could protect a watershed against floods—can result in a humanitarian crisis. Sometimes, games can help people make smart choices for the future.

Peter Perla, a mathematician and research analyst who spent much of his career at the Center for Naval Analyses, is widely seen as a leading expert on war games—used for centuries to simulate the dilemmas that arise in wars and on battlefields. The history of war games dates back thousands of years to a game played with stones designed by the Chinese general Sun Tzu and to an Indian board game, chaturanga, the forerunner of modern chess. Those ancient games, which also inspired the novelist H. G. Wells's forays into toy soldier games, primarily simulated the positions of soldiers on a battlefield. The modern war game has since evolved to encompass questions of strategy and to get people in military leadership to play with scenarios that could arise among adversaries and allies.

Perla explained to me that because games ask people to make decisions, they can create emotional and psychological stress. Such feelings tend to be absent from intellectual exercises aimed at anticipating the future, including scenario planning. This gives games the potential to have a deeper and more enduring impact on how people view possible threats and opportunities that lie ahead. We feel, not just think, when we play a game.

Games occupy a distinct sphere between fiction and reality, Perla contends, and open up the possibility for people to suspend their disbelief about certain future scenarios. A person in a game gets handed a scenario as if it's playing out in a film or literature. Because it's not presented as real, she does not start doubting its likelihood right away. And yet, unlike when seeing a movie or reading a novel, the person playing a game is not a passive observer of the story; she is engaged as an active participant and must make real decisions in this fictional world. A player also has to deal with the consequences of making decisions in the game, including backlash from other players, discord, or catastrophe.

A war game that President Clinton played in 1998, after reading the Richard Preston novel *The Cobra Event*, focused on the potential risk of a bioweapon unleashed on the American public. The White House game responded to the president's curiosity about the risk of biological terrorism. Within a month, the president had called a special cabinet meeting on bioterrorism, which led him to ask Congress to add $294 million to its counterterrorism budget.

Games have a long history of shaping military strategy and how nations prepare for war. A series of war games played by naval officers in the 1920s and 1930s, for example, helped them to anticipate many aspects of fighting the Japanese in World War II. The Naval War College hosted more than three hundred games between the two world wars, more than a hundred of which focused on strategic questions involving a war with Japan. As Steven Sloman and Philip Fernbach have pointed out, the attack on Pearl Harbor, though its timing was a surprise, was actually expected, and war with Japan was even anticipated at the time by much of the American public. In 1941, President Franklin Delano Roosevelt moved the Pacific naval fleet from its base in San Diego to Hawaii as a direct response to Japanese aggression. (No game, it seems, anticipated the kamikaze pilots.) The risks that emerged in the naval war games made scenarios of war easier to contemplate and set into motion plans for an offensive campaign in the Pacific. One naval lieutenant commander credited the war games with helping the Navy grasp the way future wars would be fought.

On a sticky summer day in 2017, I went to the Pentagon to meet with the U.S. government's experts building war games today, Adam Frost and Margaret McCown. The Department of Defense's group on war gaming was started just after World War II, and now reports directly to the chairman of the Joint Chiefs of Staff, the highest-ranking military officer in the United States. Frost and McCown told me about the games they design, to be played by military commanders and high-level officials at the White House and in federal agencies, as well as by foreign allies. Many of the games are classified, but some aspects can be described here.

McCown and Frost say that what is most important about games is they help people contemplate what has previously been unfathomable. For example, one game about the U.S. posture in the Middle East presents its players with a scenario in which the ruling Saudi king dies. All have to play their respective roles and react to this scenario with decisions. In games like this, the war game designer invariably includes a red team, or adversary, and sometimes also includes allies or neutral parties. In the middle of the game, a twist might be presented that helps people explore what decisions they would make when caught by surprise and what potential downside or upside consequences could emerge.

Frost described to me the game Pocket Century, designed at the request of the U.S. Agency for International Development (USAID), which top government officials played during the Obama administration. At the time, the U.S. government was doing a lot, through the Department of Defense and other agencies, to prevent the Mosul Dam in Iraq from failing. The dam, one of the largest in the Middle East, lies along the Tigris River, which runs from the northern border of Iraq south to Baghdad, and empties into the Persian Gulf. Control of the dam has been pivotal in the war against the Islamic State of Iraq and the Levant (ISIL, aka ISIS).

But most U.S. officials outside USAID were not considering what would happen—and what the United States would do—if the dam actually did collapse. The consequences were in some ways too devastating to fully contemplate—more than a million Iraqis were in danger of being flooded, and the humanitarian disaster could create an opportunity for ISIS to extend its influence.

Pocket Century simulated the failure of the Mosul Dam. In the game, it emerged that all the United States had been working for in Iraq would come unhinged if the dam collapsed. USAID representatives indicated in the game that they would want to send assistance to the region amid the crisis, but Pentagon officials said they would not be able to provide protection amid the chaos, and would instead need to pull out. They saw their mission not as preventing humanitarian disaster, but as defeating the enemy if feasible. Leaders realized during the war game that the dam's failure could create a total breakdown of U.S. strategy, with two major factions in the government at odds about what to do.

After the war game, Secretary of State John Kerry moved money toward managing a potential dam collapse, including dedicated funding for flood preparation and early warning systems for Iraqis in the region. President Obama put it at the top of his agenda for a meeting with leaders in Iraq. They acted to prevent crisis ahead of time.

Role-play games, whether for purposes of peace or war, seem to work best when they include scenarios that people either dismiss or do not fully believe. What games can do, however, is limited by those scenarios.

Perla says there is a danger that designers of a war game will sanitize rare, unsavory events—leaving out the collapse of the dam, for example—and therefore reinforce the bias for what players have already experienced in the past. War game designers should have no dog in the fight, he says, so that they create the game with credible scenarios already under consideration, and also some that lie outside the players' peripheral vision.

That said, no designer of scenarios can ever anticipate everything of consequence that might happen—as some extreme events just cannot be predicted. As Nobel Prize–winning economist Thomas Schelling put it, "One thing a person cannot do—no matter how rigorous his analysis or heroic his imagination—is draw up a list of things that would never occur to him." The war game experts I spoke with said that having red teams in the games that act as adversaries, with free rein to make decisions as they see fit, can help with this. Having these roles played by younger people who are not entrenched in their positions or concerned with maintaining authority can sometimes elicit wild scenarios to emerge in a game spontaneously. Like in Mattapoisett's flood project, intergenerational groups can exercise greater foresight.

Games can also fail in going overboard. If they paint future scenarios that seem too hopeless, they can rob players of the feeling of agency. A war game about a bioterrorist attack, called Dark Winter, bred paranoia and helplessness among the U.S. officials who played it in 2001, and even those merely briefed about it. Journalist Jane Mayer has documented how Vice President Dick Cheney, after watching a video of the Dark Winter war game, argued for nationwide vaccination against smallpox and then holed himself up in an underground bunker during the post-9/11 anthrax scare. The game portrayed transmission rates of the smallpox virus well beyond what is likely, and massive deaths beyond the ability of any of the players in their roles to prevent. The result was no more motivating than waking up from a dystopic nightmare.

I wondered how war games might be applied to communities preparing for climate change. I got my answer from Larry Susskind, who heads the negotiation program at Harvard Law School and is an urban planning professor at MIT. He also founded the Consensus Building Institute, through which he has designed role-play negotiations and games for thousands of people, helping forge water rights agreements among farmers and cities in California and thawing relations between the Israeli government and Arab Bedouins. A couple of years ago, I sat down with Susskind at a small table in his office at MIT, where he likes to meet with students and colleagues. On the table was a set of tiles made from colorful geodes from around the world that he keeps on hand to quell students' anxiety. They worked on me, too.

Susskind and two doctoral students designed a series of games that involved scenarios of climate change in coastal communities. They brought together city officials, urban planners, and citizens in Rotterdam, Singapore, and Boston and in four New England coastal towns to play the games in small groups. The players came from different walks of life, and had different jobs and community concerns. The researchers tracked hundreds of participants for six months of game playing, and beyond for a period of two years. They found that in both the cities and the small towns, playing the game significantly increased concern about local climate change risks and the willingness to support local action to prepare for those risks. They published the results of this research in the journal *Nature* in 2016.

And they discovered something else. People who played the climate games also experienced a notable increase in their belief that taking local action could make a difference in preparing for climate change. They felt a sense of agency in light of catastrophic predictions.

Jane McGonigal, a game designer who builds multiplayer online environments aimed at motivating people to solve real-world problems, has said, "When we play, we also have a sense of urgent optimism. We believe wholeheartedly that we are up to any challenge and we become remarkably resilient in the face of failure." She notes that people playing such games fail about 80 percent of the time, yet they keep persisting.

In playing a game, in other words, people may gain a sense of power that makes it easier to engage instead of give up. This strikes me as crucial in today's society—to stay optimistic about what we can do about the future so it feels worth fighting for today. A practice communities can use to muster up that motivation is to play more games of the future.

THE LIVING CROWD

Passing Torches Between Generations

> I am with you, you men and women of
> a generation, or ever so many
> generations hence,
> Just as you feel when you look on the
> river and sky, so I felt,
> Just as any of you is one of a living
> crowd, I was one of a crowd.
>
> —WALT WHITMAN,
> *"Crossing Brooklyn Ferry"*

What will it take for communities and societies to be less reckless, not just for the sake of tomorrow or next year, but when it comes to the choices we face that ripple across generations?

In the biting cold of a Buffalo winter, six people who barely knew one another huddled in a lakefront cabin to answer such a question. The group included a physicist, an anthropologist, an architect, a linguist, an archaeologist, and an astronomer who searched for extraterrestrial life. Each of the luminaries had been handpicked by the Sandia National Laboratories of the U.S. Department of Energy.

Their charge was to figure out a way to signal to distant future generations—the people who would live on the planet for the next ten thousand years—the danger of a site in the desert twenty-five miles east of Carlsbad, New Mexico. In casks in a subterranean cavern, the government agency planned to put to rest radioactive materials from nuclear weapons tested by the United States in the decades since the first atomic bomb, known as Trinity, had set the sky ablaze in 1945. The nuclear waste would remain toxic to humans and animals for thousands of years, begging the question of how to mark its burial site to convey its risks to people and civilizations more distant from ours than the pyramid builders in Egypt, as alien to us in time as creatures on another planet would be in space.

Maureen Kaplan joined the group in Buffalo in December 1991 as the sole woman and sole archaeologist. She had earned her doctorate at Brandeis University in Massachusetts, where for her dissertation she studied what she calls ugly pottery, traded by the ancient civilizations that thrived in Egypt, Lebanon, and Syria. The pots' lack of charisma allowed her to persuade museums to let her take samples to determine whether the clay came from the Nile riverbed or the red fields of the Levant. These were clues about who made the pottery, where they made it, and how their wares were traded.

When Kaplan graduated, however, she was broke and wanted to find a job as quickly as possible. She strayed from archaeology to take the first job she was offered, tracking lobster populations near power plants that spewed warm effluent into coastal waters. In the 1980s, she began researching nuclear waste disposal questions for the Analytic Sciences Corporation, where she could wed her newfound knack for crunching numbers for environmental research with her prior training to imagine people who had lived a long time ago.

Kaplan realized she had no idea what future generations of humanity might be like, or what markers or messages they might understand as warnings. She doubted anyone else did, either. She knew only the past—but she knew it well, and she had experienced what it was like to try to piece it together in the present. She began to think of herself as part of a future generation, one that had inherited inadvertent clues from ancient Greece and ancient Egypt. She thought about what had made her understanding of the past possible in some instances—and impossible in others. The task seemed easier than trying to predict the future.

At the Acropolis in Athens, the sculptures and buildings had cracked and eroded over time, barely withstanding the ravages of nature and the raiding of their original bronze and marble by vandals. The pyramids at Giza, by contrast, had persisted for nearly forty-five hundred years because of their grand scale; their massive stones from local quarries were hard to move and did not weather away in the desert heat. As markers on the landscape, they endured, and people still understood their purpose as graves. And yet the pyramid builders' original intentions were betrayed: Tomb raiders across centuries had stolen the treasures of the pharaohs meant to accompany them to the afterlife.

Kaplan thought of Stonehenge, which had endured more than four thousand years in a much wetter climate than the Egyptian pyramids. The sandstone and bluestone megaliths reflected the work of humans marking off a territory on the Salisbury Plain in England. No valuable metal or stone worthy of stealing and repurposing makes up Stonehenge. Yet the monument does not convey its original meaning to us, its future. No written records explain it to us, such as those that have been discovered for the pyramids and the Athenian Acropolis, and interpretations of its purpose have varied wildly over the ages.

None of these monuments could serve as the perfect model for speaking to the future. None were time capsules left for us, as far as we know. People discovered them by accident.

Kaplan knew that the markers for the nuclear waste site would have to endure the changing climate over ten thousand years, the forces of wind and water and ice that level mountains and carve canyons, and the unpredictable acts of generations of wayward human beings. And unlike Stonehenge, the markers would have to offer a comprehensible message to people who might be as different from us as we are from those who carved into cuneiform in ancient Sumerian—the oldest known written language, which resembles none of those used today. "We wanted to give them a warning, but not overstate it," Kaplan reflected recently when we spoke by phone. "You don't want to say, 'Touch one stone and you'll die.' Inevitably someone will touch it, and when they don't croak, you'll have lost all credibility."

The question of how to mark the grave sites of nuclear waste for future generations was not an issue for the thousands of generations of people who lived before the mid–twentieth century. It arose because of humanity's growing knowledge and engineering prowess. With our technologies, we can shape the very survival of people to come. Even when we don't know the precise consequences of our actions, we know their longevity.

Nuclear waste is among the most enduring of our imprints on Earth, but other choices we make present similar conundrums. Today we are changing the planet's climate for future generations with the pollution from our everyday use of fossil fuels as energy. We are already experiencing climate disasters around the world owed to past failures to act, with the Arctic's melting and sea-level rise outpacing scientists' expectations. Even if we abruptly stopped polluting the atmosphere next week, our past emissions would continue to heat up the atmosphere for at least four decades. We have the choice today to put mirrors into near-Earth orbit or pump aerosols into the stratosphere to block out the sun—what's known as geoengineering—or to try to reverse the planet's warming trend by scaling up solar and wind power plants and efficient energy technology.

Scientists of our generation have the ability to edit the genetic code of human embryos with a technology called CRISPR. It can be used to delete mutations that cause genetic diseases or to create those that instill certain traits—perhaps even hair color or athletic ability or IQ. But those gene edits, if made to heritable traits of embryos, could irrevocably change the course of human evolution for the generations to follow ours, by inserting into the gene pool attributes that would be passed on. We don't fully understand the consequences downstream, despite having this capability. Most of the genome is a vast uncharted terrain. Of the small fraction of what we know of human genetic mutations, we already find trade-offs: The same mutations of the gene CCR5 that can protect people from HIV also increase the risk of contracting West Nile virus. The same genetic mutations that cause sickle-cell anemia protect people from dying of malaria.

With the Promethean power to make decisions today that will define future generations—whether it's creating their future climate or changing the species itself—we have unprecedented obligations. Yet for the most part, we lack the ability to think and plan across the spans of time in which we are implicated.

Most people do not let their minds wander beyond a generation or two into the future. This makes sense because our emotional ties extend at most to our children, grandchildren, nieces, and nephews. It might even be unreasonable to expect ourselves to have great concern and understanding of people we have never met, whether distant in geography or time. News reports and travel, at least, bring us closer to people from far-flung places on the planet who are suffering; we get no such flashes of those distant from us in time.

Kaplan's group that gathered in Buffalo in 1991 eventually joined a second group in New Mexico to trade ideas. Together, the experts concluded that the desert waste site should be marked with jagged-edged obelisks and warnings in multiple languages, as well as pictures of faces bearing expressions of horror. Yet even the experts acknowledged the futility of their own effort. They concluded that after five hundred years, the likelihood of people being effectively warned by the messages would drop steeply. Maybe civilization, at least as they knew it, would no longer exist.

There was no way to be confident that a signal flare would be heeded in the distant future.

The group that came together in the early 1990s was not the first to contemplate the question of how to convey the risks of our generation's nuclear waste to distant generations of the future—nor was it the last. In 1980, the U.S. government had asked several technical experts, including Maureen Kaplan, for ideas about marking future disposal sites for the commercial waste from nuclear power plants. Wacky ideas emerged, including one to genetically engineer cats whose fur would turn green in the presence of radiation—under the assumption that the long-standing companionship between humans and cats would persist far into the future. Other ideas were callow, such as making signs with skulls and crossbones to signal danger. Even today, this symbol might convey poison or a pirate theme park.

Linguist Thomas Sebeok argued in 1981 that the nuclear waste problem required an atomic priesthood, a Freemason-style group of sages who would pass along the knowledge of waste sites to subsequent generations—to account for the fact that over ten millennia a message was likely to get lost or lost in translation. He also proposed seeding world religions with myths, passed down in oral tradition, describing the dangers of nuclear waste. None of the ideas has been taken up as of this writing.

Decades later, in 2014, another group of scientists, historians, artists, and anthropologists gathered in Verdun, France, to tackle the same question. Their gathering was organized by the nuclear energy arm of the Organisation for Economic Co-operation and Development (OECD), a forum composed of the governments of more than thirty democratic nations, and they called it the "Constructing Memory" conference.

While the ideas that emerged from each of these groups are fascinating thought experiments, it is unclear whether any will ever be enacted. The cabals may have been useful in activating people's imagination about the future—similar to the ten-thousand-year clock—but as practical plans, they faltered.

Meanwhile, engineers in Finland have been constructing the world's first long-term storage site for commercial nuclear waste, where they plan to start burying radioactive material in 2024. The spent fuel from power plant reactors—as opposed to the waste from weapons tests being buried in New Mexico—could remain dangerous to humans and animals for up to a million years. An analogous effort in the United States to build such a site in Yucca Mountain, Nevada, has stalled in part because of public opposition and the maneuvering of powerful opponents such as former U.S. Senator Harry Reid.

The architects of the Scandinavian site Onkalo—Finnish for "hiding place"—are hoping the forest will grow over the entrance to the underground tunnel. The idea that the waste site will simply be forgotten by humans seems naive when you think about trespassers across a million years. Today we can survey the surface of Earth at granular scale with satellites and drones and see the imprints of human disturbance on landscapes. Ocean explorers have already stumbled upon lost shipwrecks from mere hundreds of years ago.

That people around the world have tried to solve the nuclear waste dilemma for decades reflects that collectively, we feel responsibility toward the future, at least for the extent of the known life span of our technologies and toxins. Time and again, however, people have stumbled upon the futility of making plans for a far-off time, and for good reason.

Early cave-dwelling humans of the Paleolithic era could not foresee the farms and civilizations that would spring up along the Tigris and Euphrates rivers thousands of years later. The ancient Greeks who drew up democracy did not know that their ideas would feed revolutions against monarchs centuries later. The textile workers of nineteenth-century England did not imagine that in just a hundred years or so, their descendants would mostly buy not the goods of nearby factories, but shoes and T-shirts made in China, shipped across oceans. Their descendants would board planes that flew in the stratosphere from Tokyo to Los Angeles.

It is only getting harder to envision what future generations will experience. Our inventions are driving profound change in society in each successive generation, and even within the span of lifetimes. When my parents immigrated to the United States, they sent letters that took three weeks to reach their families in India. They never imagined a world with the internet, the smartphone, instant messaging across the planet, social media networks, or GPS navigation. Just ten years ago, I could not have foreseen an autonomous drone delivering a package to my front door, a 3-D printer creating a working handgun and a human artery, or a robot that could drive a car in traffic. Someday soon, other technologies we can't predict today will make the world you and I have known unrecognizable.

This was not always the case—that each generation in human history found the world inhabited by the immediate next generation to be unfathomable. In fact, for a long span of history, many of the basic features of human existence remained fairly static. While it took ten thousand years from the dawn of civilization to the invention of the airplane, it took less than seven decades from the first airplane flight in Earth's atmosphere to the moon landing, notes the science writer Michael Shermer. In a matter of decades, the world as people know it today will seem distant to us. The amount of progress and disruption that took twenty thousand years in human history, Ray Kurzweil asserts in his seminal 2001 essay "The Law of Accelerating Returns," will henceforth take only a century.

It's not just technology that is changing faster than in previous eras of human evolution; it is society itself. Shermer has charted the average duration of sixty civilizations across history, ranging from ancient Sumer and Babylonia to the eight dynasties of Egypt to the Roman Empire, and including the dynasties and republics of China and modern states in Europe, Africa, Asia, and the Americas. He finds that the life span of these civilizations—on average 421 years—has been decreasing since the fall of Rome.

The collective discontent that arises from rapid technological and social change got a new name in the 1960s. Alvin Toffler called it "future shock"—a kind of cultural disorientation that we experience in our own society. In his 1970 book named for the syndrome, Toffler argued that the mass affliction creates paralysis when it comes to exercising wisdom about the future.

Looking far into the future can feel futile for philosophical reasons as well. At some point, an asteroid might destroy Earth, or nuclear

Armageddon may render our current decisions about the future moot. Or as the economist John Maynard Keynes famously quipped, "In the long run, we're all dead."

If you prefer the chaos theory variant of nihilism, you might consider that a butterfly flapping its wings in Chicago might eventually—and unpredictably—trigger a hurricane on the other side of the globe. Looking back, someone might have shot Hitler's grandparents and spared millions of people's suffering and death, or thrown themselves in the line of fire of Archduke Ferdinand and prevented World War I. But how could they have known?

We cannot predict today how any decision, big or small, might trigger a chain of events with extraordinary effects on humanity's future. Unwieldy consequences cannot possibly be taken into account in decisions we make today—but that does not mean we should ignore the future consequences of all decisions. Some scenarios, like an asteroid destroying Earth, are highly unlikely in the foreseeable future, while others—like rising seas—are near certainties.

Is it okay to let ourselves off the hook? Can we just forget future generations in our decisions, whether to leave nuclear waste around unmarked, or to heat up Earth's climate with abandon?

We are saying yes by default to this question in our generation, without asking ourselves the question directly. The way many democratic societies have made decisions over the past few decades has neglected the interests of the young and of people not yet born. One prevalent tool of traditional economics has served to justify that disregard.

I came across this tool when I was a graduate student at Harvard a decade ago. While I attended classes by day, I worked Friday and Saturday nights on the metro desk of *The Boston Globe*, tuning a police scanner so that I could report on late-night house fires, car accidents, and arrests. I loved the contrast between the immediacy of the newsroom, where my job was to cover the news as it broke, and the intellectual life of grad school, where I reflected on trends and studied the underlying drivers of the global economy.

One Monday morning, after a particularly sleepless weekend, I found myself blinking to stay awake in class when the professor introduced a concept that roused me from my quasi-hibernation. It was the first time I heard about what economists call "social discounting."

To explain the concept, the professor pointed out that it is rational for people to value rewards that they expect to arise in the future somewhat less than they value those same rewards today. It was a wonky way of saying that a bird in the hand is better than two in the bush—it takes time to get those two birds out of the bush. It made sense to me, as long as people did not overly weigh the immediate over the delayed, as we so often do.

The professor then turned to telling us how this concept of discounting is used by governments making policy—indeed by entire societies. For example, when an agency decides whether it is worth investing in a long-lived bridge, or creating a national wildlife refuge, officials evaluate its life span and the benefits and risks of that project over time. Rewards that come in the distant future are discounted by a rate that bureaucrats deem appropriate for the market rate of return—in other words, how much the pot of money would be worth over that same time span if you chose to invest it in the stock market instead of the project.

As I jotted down notes, I found myself shaking my head in disbelief. In the halls of academia, the approach seemed logical. But in the real world, it was at best incomplete—and at worst a deeply flawed way to reflect a society's values and aspirations.

It's one thing to discount the future for yourself—based on your own preference to get some money now instead of more money next month. But how could governments choose a single number that reflects the value that all citizens or communities place on getting future rewards or escaping future dangers? And how could they even begin to imagine how much future people would value clean water and air? We know that people regard the future differently, depending on how desperate or sated they feel, their environment, their culture, and their circumstances. Future generations are not around to say how much they value what we are leaving behind or robbing them of today. They have no voice in present deliberations.

Even within a single generation, there are countless cases where invested money cannot stand in for future benefits. The government funding that supported basic research that led to the internet and GPS would not have reaped anywhere near those projects' eventual returns to society if it was instead invested in a savings account or the stock market. Invested money has never fully replaced valuable natural resources lost after a catastrophic tipping point, such as the now bereft New England cod fishery or the dying Great Barrier Reef. No amount of money could substitute for the Sistine Chapel ceiling or van Gogh's *Starry Night*.

Social discounting makes it possible to undervalue these resources, making them interchangeable with the amount of money that current generations are willing to spend on them in a hypothetical future that they may never see. Using such practices, no country today would build the Brooklyn Bridge, the National Park Service, or the Great Wall of China. It's little wonder that the same societies do far too little about problems like climate change that will profoundly impact future generations.

Consider a hypothetical question: Would you agree to let a million people die sometime in the future to save one person's life today? Or how about sacrificing the lives of thirty-nine billion people a few centuries from now to prevent one death today? Tyler Cowen, the economist and ethicist, points out that most people would say no to these questions from a moral point of view, and yet using any social discount rate above zero essentially means we say yes to these questions, weighing the present generation disproportionately over future generations. Cowen is among the economists who now see the limits—and even the folly—of social discounting given our era's legacies and our knowledge, which require new approaches when we are dealing with distant generations.

Social discounting has become a dominant practice of Western democratic societies when it comes to weighing future consequences of collective decisions. Experts rely on a social discount rate to determine how much to weigh what happens in the future—to us and to our children and grandchildren—to tell us whether it's worth it to act today, and how drastic our actions should be. A high rate means we place less weight on what happens in the future, and a low rate means we place greater weight on it. The number economists choose for their calculations—a topic of considerable debate—makes all the difference in whether we see programs or policies as destroying our economy or saving it. But the public does not scrutinize these rates, and for the most part, the rates do not reflect the values of most people.

Social discounting is not entirely responsible for societies' disregard of future generations—greedy and weak leaders can play an outsize role, as can the voters who don't hold them accountable. But the tool is emblematic of how political and economic systems fail to value consequences across time, and it can serve to justify and obscure leaders' myopic decisions.

When it comes to climate change, the social discount rate has been a hidden, critical factor in how far political leaders and the voting public are willing to go to tackle the problem. Is it worth it to take action today on climate change to prevent humanitarian disasters tomorrow? How much is worth doing? If we put a price on carbon through a tax or a trading scheme, how high should it be? Economists and policy makers considering these questions account for the costs of investing in new jobs for coal miners, rebuilding roads and electrical grids, using less energy, and scaling up cleaner electricity sources like wind and solar as well as technologies for energy storage and transmission. They account for the costs of seawalls and of moving communities from vulnerable locations. And they try to measure the unknown costs to society of a devastatingly warmer world in the future—a world of climate refugees, destroyed farms and forests, and flooded cities. But they never weigh the future consequences quite enough.

Deep concern for future generations is a near-universal human value, one that has echoed across cultural and political boundaries, and across the ages. Despite current practice, to devalue future generations is at odds with the founding principles of democracies, not to mention major world religions and the moral codes of atheists and agnostics.

The eighteenth-century Irish political philosopher Edmund Burke, hailed as the godfather of conservatism, described society as a partnership among generations. The partnership is "not only between those who are living, but between those who are living, those are who are dead, and those who are to be born," he wrote in 1790.

Georgetown law professor Edith Brown Weiss has argued that Burke's partnership is a social contract between a government and citizens to "realize and protect the welfare and well-being of every generation in relation to the planet." In her seminal scholarship on intergenerational equity, Weiss notes that the idea that every generation should steward the planet's resources for its descendants is a concept that "strikes a deep chord with all cultures, religions and nationalities."

The English political philosopher John Locke drew on Judeo-Christian teachings to argue that humans could use only as much of the world's patrimony as left "enough, and as good" for others. The strain of thought has carried over, at least in theory, to common and civil law influenced by English tradition around the world.

Thomas Jefferson wrote to James Madison of an equal entitlement of every generation to Earth's resources, "unencumbered by their predecessors." Theodore Roosevelt spoke of the duties to unborn generations that required less waste of our heritage from the "present-day minority" of humanity.

The public trust doctrine, which holds that certain cultural and natural resources should be protected by governments on behalf of current and future generations, dates back to ancient Roman and Byzantine law, and has been carried forth in the language of modern democratic constitutions. Legal scholar Michael Blumm has documented how this doctrine has been drawn upon in judicial decisions in a dozen countries in Africa, South Asia, and South and North America. Customary laws of various communities in Africa consider living people to be tenants on Earth, with obligations to people past and future. Customary land law in Ghana, for example, posits that land is owned by a community that transcends generations. The Constitution of the Iroquois nation stipulates that decisions be made with an eye to the consequences for unborn generations.

The idea that we should act in deference to future generations also permeates religious belief systems spanning Christianity, Hinduism, Islam, Judaism, and Shintoism. It echoes in the words of history's most influential poets, from Walt Whitman to Pablo Neruda, from Rabindranath Tagore to T. S. Eliot.

The problem is that while valuing future generations is an ideal embraced across cultures and ideologies, it is not yet a widespread cultural or institutional practice. Our shared principles have not held sway in reality, leaving a vacuum that has been filled by social discounting and other practices at odds with our concerns and obligations to the future.

We need a different way to think and act on our concern for generations to come in this age of immediacy—one that positions us neither as hapless daydreamers breeding glowing cats nor as callous thieves discounting future catastrophe. We need to think and act as ancestors.

A few years ago, I visited my grandmother in India. Memories flooded back of childhood trips to see her, of my sister and me pinning jasmine garlands in our hair and filling a bucket to pour water over our sticky bodies in the bath. For as long as I can remember, all have been welcome to pass through the wrought-iron gate into her home: gossiping neighbors, whiny street cats, and droves of mosquitoes with a taste for American-born blood. Each morning, she rose before dawn to sear bitter gourd and pop mustard seeds over a flame.

On this recent visit, she had something she wanted to give me. She directed me to climb onto a rickety chair in her bedroom and reach above her armoire to bring down an heirloom that she had saved for me. We sat together on the mattress, hard as a coffin, as I unraveled the fraying rope and unfurled the batik cloak that clung to the antique.

The heirloom belonged to my great-grandfather, a music and art critic named K. V. Ramachandran. In letters he exchanged with a UCLA composer and professor known as the father of ethnomusicology, Colin McPhee, he defined for Western ears the rhythms of Indian drumming. Occasionally, he heaped praise on performers, weaving a tale of a mongoose that turned to gold after hearing the voice of a classical singer. The instrument I held in my hands, called a dilruba, had been custom-made for Ramachandran as a gift, and its twenty-two strings had once reverberated with a sound that aroused the image of a melancholy wanderer in the Himalayan fog.

I did not know this man, my great-grandfather, although I had read some of his criticism. I knew him no more, in a sense, than I knew Walter Cronkite from seeing his face in old footage of news broadcasts, or Vincent van Gogh from seeing his paintings in museums. I only vaguely knew K. V. Ramachandran from my grandmother's stories, as the person she had most admired. He insisted on her ongoing education, in a time and place where girls were given away as child brides in their early teens. He taught her to sing, and hired a teacher who taught her the classical storytelling dance known as Bharatnatyam, an art form she perfected to the point of later being invited to dance for the Sultan of Brunei and the Queen of England.

The heirloom sits with gleaming new strings in my living room today. Its connection to my great-grandfather's memory and its ageless beauty haunt me. To trace its wooden carvings and inlaid pearl with my fingertips, to pluck its strings, is to be linked to an unknown time and place. I had never really thought about my great-grandfather until I had something in my possession that linked me to him, that made him feel present. Now I think about him often, about an ancestral home that I never inhabited, about a past I don't know. It is my own version of Scrooge's ghost of the past.

While the instrument is in my possession today, I know that it doesn't actually belong to me—it belongs to the generations of my family that came before, and to the ones that will come after. It belongs to time itself.

Within a family, an heirloom transmits traditions, values, and stories of the past. It carries with it the notion that future generations matter to the present generation, and that past generations will matter to the future. Some heirlooms are saved out of vanity or nostalgia, yet still embody the concern we have for what happens over time. Each generation serves as an heirloom's steward, imbuing it with its own meaning, while carrying forth its tradition. My grandmother never played the instrument she gave me, the dilruba, but she never threw it away, either: She knew it was a sacred object even when she could no longer retrieve it from its hiding place in her old age. It is my most treasured belonging, not because I knew its original owner or his full intentions, but because it gives me meaning. The heirloom designates me with a purpose, to usher it along in time.

Heirlooms position us as both ancestors and descendants, a sharp contrast with how we tend to think of our communities and our society today. The experience of touching something timeless can shock a person out of the rhythm of ephemeral experience in daily life. I have wondered whether it is possible to orient ourselves to this way of thinking and acting not just when it comes to our own family heirlooms, but to the collective problems that face our society. What would it look like?

For one, if we thought of ourselves as keepers of shared heirlooms, we might focus more attention on what we pass to proximate generations—but not look much farther down the road than the next two to three generations. This would make what we pass along more connected to people we know and love—our children, grandchildren, students, and godchildren. Collective heirlooms would not be time capsules buried for people ten thousand years in the future to open, in other words. Each generation would instill the immediate next with the knowledge it needs to use and pass on the inheritance. These shared heirlooms might be ideal for situations when we can't see what the distant future will need or be like, but know there is something valuable to pass along in time—a natural resource, a piece of cultural heritage, or scientific knowledge, for example.

With an heirloom, each generation is both a steward and a user. The heirloom doesn't have to sit on the shelf untouched, in other words, but can be used by each generation as long as the use does not deplete its benefits for the next. Legal scholar Mary Wood, drawing on Edith Brown Weiss's idea of intergenerational equity, has proposed a way that principle might work in practice for shared natural resources such as clean water and forests. Wood thinks we should treat such resources as a trust, with each generation serving as both trustees and beneficiaries, codifying our roles into laws and treaties. As co-beneficiaries, we share what lies in the trust—minerals, forests, ocean life—with others alive around the world, which creates the obligation to use it equitably within a generation.

When we pass on an heirloom, we don't prescribe what each steward must do with it. Instead, we leave options open to the next generation— like jazz standards that can be performed differently by successive groups of musicians. Keeping shared heirlooms might mean leaving a diverse range of choices to the future, and as much knowledge as we can about what we are leaving. This would keep us from having to guess or dictate exactly what future generations will want to do with what we leave them.

Societies already keep some collective heirlooms. When we protect elements of cultural heritage like Petra, the Taj Mahal, and the *Mona Lisa*, we do so because of what they represent to us about the history of humanity, and for the sake of the future of it. Heirloom keeping, you might argue, is the mind-set that has allowed National Parks in the United States to persist for more than a century, despite increasing demands for land and resources and ongoing public controversies about protected species like wolves that thrive near parklands.

In 2017, I went to see Harvard economist Richard Zeckhauser at his office in Cambridge overlooking the Charles River. Pictures of his two granddaughters decorated his door. Zeckhauser, in his late seventies, is hardly a bleeding heart; early in his career he was one of Secretary of Defense Robert McNamara's "whiz kids," who helped forge Cold War military strategy. Zeckhauser told me that in looking across human history, he felt it was safe to assume that future generations would be wealthier than ours. But in his view, this means we owe generations to come *more* cultural and natural resources—not fewer. His two granddaughters, he said, might value the Chinese porcelains at the Museum of Fine Arts in Boston more than even we value them today, especially if they are wealthier than his generation. Given their heritage and distinctiveness, the porcelains could not be replaced by mere money.

Zeckhauser and his colleague (and former secretary of the Treasury) Larry Summers think that the principle of stewarding an inheritance can be extrapolated to society's choices that affect the future of humanity. An effort in the current generation, in their words, can "set a precedent for generosity" in future generations, creating a chain reaction, if framed the right way at its inception and as it is passed along.

Zeckhauser and Summers advocate more government programs that imbue each generation with a sense of what they have inherited and what their responsibility is toward the immediate next generation. With more time and years of tradition, what we inherit feels more sacred, and more like an endowment to keep perpetuating into the future. As an example, Zeckhauser points out that turning the Grand Canyon or Yellowstone into office parks would be a lot harder to pass by the public than President Trump's 2017 decision to reduce protection for two lesser known, more recently established national monuments, Bears Ears and Grand Staircase-Escalante in Utah.

In my own work, I have witnessed firsthand how communities can come to treat a resource as a shared heirloom.

Eleven years ago, I traveled to the far western reaches of Mexico's

Baja Peninsula, where fishing villages speck the coastline—outposts that at the time were hundreds of miles from hospitals, paved roads, and cities. Motorboats departed at dawn from their harbors for the wide blue frontiers of the Pacific Ocean.

I spent several weeks on boats and on land with lobstermen from nine villages there, to learn how their fishery had become one of the most lauded in the world. They had protected and rejuvenated the Pacific red rock lobster fishery at the same time that their counterparts in the Yucatán Peninsula were decimating lobster populations and trawlers in Mexico's waters were killing dolphins with their seine nets while sweeping for tuna. International auditors had been coming to the Baja fishery for years to gauge the health of the lobster population before I went there.

The lobstermen across the villages I visited learned their trade from their fathers and uncles and are passing it down to their sons and grandsons. The villages are organized into fishing cooperatives that date back to the 1940s. A crash in the local abalone population in the 1980s caused by erratic El Niño weather spurred local leaders to band together to protect the lobster fishery like a common heirloom, as a resource and way of life that could be passed down to their sons, and their sons after that. Within the lifetime of the elder fishermen in these villages, they had transformed from hunter-gatherer ways of catching lobster to farming the sea, collaborating across the communities and with generations yet to be born.

The fishermen from the nine cooperatives plant their bait in traps on the seafloor and mark them with buoys color-coded for each fishing team. They return a few days later to crank up the cages with pulleys, using calipers to measure their catch. They methodically throw back lobsters that are too big and could be strong brooding stock, or too small and needing time to grow. The Pacific lobster has no front claws, like its cousins in Maine, but brandishes a voluptuous tail, whose meat simmers with spices in the taquerías and backyard grills on the peninsula.

The cooperatives together form a regional federation that polices the fishery to keep each village and fishing group honest and to protect the fishery from poachers. They do not rely on government enforcement of their fishing concessions. The cooperatives own the boats and fishing gear, painted and marked clearly to denote their permission to fish for lobster along this coast. The federation divvies up each season's allowed catch among the fishing cooperatives. They have essentially extrapolated the concept of the family heirloom to steward the lobster across communities.

Theirs is a distinctive fish story. Countless fishermen around the world—from Africa's Lake Victoria coastal communities to Mediterranean pirates stalking the bluefin tuna—have profited in the present by depriving the next generation, killing off fish stocks critical to the diets and livelihoods of millions of people.

On the far reaches of the Baja Peninsula, there is very little economic activity other than fishing, which makes the future of the lobster catch essential to the future of the villages and their families. Homes and businesses in these communities still ran on generators at the turn of the twenty-first century; the electrical grid did not extend to them until 2005. In recent years, some former fishermen and their families have begun to earn a living in the emergent tourism industry that has sprung up around the Vizcaíno Biosphere Reserve, whose warm waters attract breeding gray whales each winter.

Intimacy helps the cooperatives in Baja California Sur steward their shared heirloom. The nine fishing communities are small, and not far from one another geographically. They benefit from a continuity of culture, punctuated by friendly rivalries, and over time, cultural continuity has been abetted by the peninsula's isolation from mainland Mexico. The relatively limited range of the lobster makes it possible to manage among people who share similar values and the same future, even if not identical community conditions. (It also helps that the shellfish crawl into traps, lured by bait, and survive being surfaced, measured with calipers, and thrown back into the ocean—not the case for all types of fish.)

The lobstermen and families across these communities know that the fate of the fishery is their shared future—much like all of humanity will share the future state of the planet. But in the case of communities in close proximity and of small size, the people also frequently interact with each other. They know they will continue to interact and negotiate with each other over time.

I have learned of other shared heirlooms from people I have met while researching this book. Stewart Brand, the futurist and technologist who has inspired many of Silicon Valley's leading inventors and the modern environmental movement, told me when we spoke in 2016 about the Ise Shrine in Japan, a Shinto temple that was first built in 4 BCE. For more than a thousand years, people have rebuilt the wooden shrine on a twenty-year cycle, creating a perfect replica next to the existing building and then dismantling it to repurpose the wood for other shrines. Brand calls this a "living monument" because it has been perpetually renewed. It is also, in my view, a collective heirloom, shepherded by successive generations whose belief in Shinto Buddhism imbues them with the responsibility to keep up the tradition. The ritual also passes down the architectural skills of shrine building to each successive generation.

A secular version of collective heirloom keeping has protected a forest in the Italian Alps known as the Bosco Che Suona, or, loosely translated, the woods that sing. The spruce trees in this forest were first discovered by Antonio Stradivari and his colleagues more than three centuries ago, and their wood was used to fashion Stradivarius violins. The wood is believed by many musicians and craftsmen to have qualities that make for the most melodious instruments in the world. The careful harvest by luthiers of spruce trees from the forest helps clear the way for sunlight to reach small saplings that rejuvenate the forest. The cultural heritage of the forest and demand for its wood across generations have prevented it from being either clear-cut or not harvested at all, and therefore not replenished. It is treated as a sacred object, not readily exchangeable for money across generations.

In each of these instances, a community's size, or at least its cultural continuity between past and present, has made it easier to create and steward collective heirlooms. Similarly, of the hundreds of stone markers dedicated to past tsunamis in Japan, the two that were heeded centuries later were both in small villages, where oral tradition and school education reinforced the history and passed down the warning over time.

Across democratic societies, we see great diversity of family and cultural histories and interests. Societies today change rapidly, thanks to their integration into global markets and their exposure to technology. At this scale, it is difficult to imagine shared heirlooms springing up often. Even if a small group chooses to become stewards of a resource, it may lose out if others neglect or deplete that same resource, playing out the so-called tragedy of the commons. This makes it unlikely that we will see enough serendipitous shared heirlooms crop up for this generation to meet its aspirations for the future.

So how can we come to see ourselves as both ancestors and descendants in today's advanced societies? We will need practices that cajole decision makers to heed consequences for the future of humanity. Wise policy might encourage the protection of important shared heirlooms amid greed and periods of neglect. Laws and government programs could compel the reckless—certain political leaders and publicly traded corporations—to serve the role of stewards. Institutions with longevity—universities, libraries, philanthropies, churches, and synagogues—can also play a role. Using this architecture could create barriers for those who in any particular era might prefer short-term gain over stewardship of shared heirlooms.

The U.S. National Park Service and the UNESCO World Heritage sites are examples of how collective heirloom keeping can be codified into law and norms and sustained through organizations' support. Parks and the heritage sites—which can't easily be paved over for shopping malls—remain a narrow exception today to the way societies treat their inheritance from previous generations.

How can we become better ancestors when it comes to the choices that put the future of humanity at risk?

Georgetown legal scholar Edith Brown Weiss thinks that intergenerational equity should be enforced more aggressively when governments and companies make decisions about natural resources and cultural heritage, so that future generations are consistently taken into account. When Weiss first proposed this idea in her book *In Fairness to Future Generations* in 1988, she recommended that the United Nations designate a high commissioner for future generations; this has yet to happen. Nevertheless, in recent years there are signs that the principle of equity across generations is at last gaining traction, even if not at the level of global deliberation.

Several judicial decisions and lawsuits that have emerged around the world show the potential of enshrining the idea of intergenerational equity into law. At least twenty countries, Weiss says, have had courts make the interests of future generations relevant to their decisions. In some instances, courts have given legal standing to children as representatives of future generations, including in a seminal decision in the Philippines in the 1990s in which the country's Supreme Court banned the granting of new timber licenses for rain forest clearing on the grounds that it infringed upon the rights of young and future generations to a "balanced and healthful ecology." In the United States, a groundbreaking lawsuit is making its way through the courts as of this writing. Its goal is to hold the federal government accountable for the harm imposed on young people and future generations from burning fossil fuel energy, including the leasing of federal lands for oil, gas, and coal development. In 2016, a federal judge in Oregon gave standing to young people to bring the lawsuit against the executive branch, on the grounds that their constitutional rights could be violated by the government's contributions to climate change.

The Supreme Court of India has made two decisions in the twenty-first century to prevent deforestation and to protect historic water reservoirs, citing the rights of future generations. In two coal-mining cases also in India, the country's Supreme Court limited the amount of mining that could be carried out in the state of Goa, and granted a permit to mine only on the condition that the company create a trust fund for future generations to compensate them for the harm from the mining. In Brazil, the High Court handed down several decisions between 2007 and 2011 to protect the environment, citing a legal duty to future generations. The International Court of Justice has also begun to consider obligations to future generations in its opinions.

Another way governments can provide voice and weight to the interests of future generations is by designating ombudsmen to represent them in national governments. The legendary ocean explorer Jacques Cousteau inspired the French government to establish a council on the rights of future generations in the early 1990s, of which he served as the chair. (As a result of controversies over nuclear weapons testing in the Pacific, the council was short-lived.) Finland has a committee of seventeen members of its Parliament who represent future generations, and for six years, Israel's Knesset had a Commission for Future Generations, which looked at bills before the legislature relevant to science, health, education, technology, and natural resources to evaluate their consequences for future generations and to recommend ways to make the policies more farsighted. Hungary, Germany, and Wales have all had similar positions for ombudsmen at various points. Politics has gotten in the way of some of these efforts. The trick is to insulate such positions from political maneuvering and to ensure that the people who are appointed are credible and steadfast in carrying out their duties on behalf of the future, not current special interests.

What would it mean to treat the problem of nuclear waste as if we were keepers of shared heirlooms? I think it would change how we look at the problem. For one, we would not concern ourselves so much with creating a time capsule, monument, or message to communicate with people alive fifty generations from now—as intriguing as it is to indulge that fantasy. Nor would we hide the waste and hope it is forgotten, as the

Finns are doing. In the tradition of heirloom keeping, we'd focus on prox-imate generations—on giving them knowledge and options, and on not leaving them worse off than we have been. And we'd try to instill in them a sense of responsibility to each next generation.

To do right by the next generation, we would focus attention not on marking waste sites, but on coming up with a plan for the waste that is now temporarily left on-site at nuclear power plants, in storage pools or casks. Since plans to build a site for long-term nuclear waste storage in the United States have stalled because of political opposition, we may need to create a group of ombudsmen representing future generations that would work with private companies and government and communities to plan for the waste.

To make sure that future generations have as many options as possi-ble, we should not bury nuclear waste irretrievably—for we can't anti-cipate whether people might one day find uses for it that we cannot imagine. Some optimists envision a future where people might use nu-clear waste silos for heating. Writer Juliet Lapidos speculates that future humans, if medical progress continues, might find cures for radiation sickness.

We might also teach the next generation about nuclear waste, its risks, and the assumptions we are making in how we store it today. Long-lived institutions—universities, libraries, museums—could be called upon to convey dangers and potential uses of the waste. We might start to teach school children about nuclear waste and where it is buried on the planet.

Every object we excavate from our parents' basements cannot become a family heirloom. Similarly, we cannot make everything in our midst a collective heirloom. It would not be practical to steward every resource, artifact, or investment a society makes across generations. But we might decide that the most important resources deserve the architecture of shared heirlooms. And we might call upon the practice of heirloom keeping when we are faced with choices today likely to severely impact future generations. When the stakes are high—or potentially permanent— we have greater need for heirloom thinking.

We may not know what people in the future will wear, how they will travel, or what devices will be implanted in their brains. But we do know that people, however long we persist into the future, will retain something of what it is to be essentially human: the drive for survival, and the need for natural and cultural resources to support that survival. The pursuit of pleasure, knowledge, love, beauty, and community. We know they will seek a sense of belonging to time, as the ancients did and as we do today—one that connects them with the past and with the future.

When it comes to the gene editing of embryos, looking at our choices as ancestors and descendants, rather than as those of people isolated in time, might lead us to different decisions. Human genetic diversity is our inheritance, a source of resilience in the face of disease and other threats. Stewarding the diversity in our genomes creates options for future generations to draw upon for understanding themselves and the scourges of their own generation. To leave the most options for the future, we might collect as much knowledge as possible about the genome, and about the potential trade-offs in making edits to it, but avoid engineering human embryos in ways that diminish genetic diversity or that prevent later generations from gaining knowledge about us, their ancestors. We can use the technology to fight disease in the living without tampering with the genetic pool.

The climate crisis is likely to define today's generations of humanity as either those that made a giant leap in foresight or the last ancestors at the party. It is the preeminent problem that calls for us to rethink both the conventions of social discounting and the uselessness of time capsules: We have a high degree of certainty of the dangers to both present and future generations from society's current course of action, including threats to social order and public health, rising seas, and catastrophic damage to Earth's landscapes, oceans, freshwater, and biodiversity. We also know what we could do today—cut emissions and prepare for disasters—to reduce the harm to future generations, but the actions may seem too costly under the calculus of discounting. In our generation, we are reaching points of no return in warming the planet that would make it meaningless to merely warn the future about what we have done.

We have the power to change how we look at this problem from one of avoiding economic loss to one of stewarding collective heirlooms—the atmosphere, the oceans, and the diverse landscapes of the planet might each be seen as resources to be shepherded across time, to be used without being destroyed irreparably. Iconic cities, forests, rivers, prairies, and beaches could also be seen as cultural heritage to be preserved as heirlooms. We would need ways to understand costs and benefits, and to enforce this way of seeing our resources so that societies take action to avert the worst of climate change. Mary Wood believes we can do this by establishing trusts for natural resources, whereby communities, companies, and governments have the obligation to leave intact the "principal" of what we have been endowed while using what we can glean of nature's capital.

To leave more options to future generations in the spirit of heirloom keeping, we would invest far more heavily in scientific research into renewable and cleaner forms of energy, and new ways of transporting ourselves and organizing our cities and communities, so that we can pass that technology on to future generations. We might even pursue more research schemes to cool the planet, leaving the future the possibility to use that knowledge.

In my view, passing shared heirlooms will require us to have more contact and connection between young and old, not just in our families but in communities and society. This can take many forms. Governments at the local, state, and national level can create venues and channels for youth to participate in government, before they are even of voting age—as advisers to city councils or ministries, for example. Churches, temples, mosques, and synagogues can create intergenerational groups, instead of segregating youth groups from the middle-aged and elderly. Businesses and organizations can designate board members to represent the interests of future generations and consult youth not just as consumers but as idea generators and advisers on values. The young can lead the old, as the high school students who survived a horrific shooting in Parkland, Florida, did in 2018, when they sparked a national movement of teenagers asking parents to sign contracts pledging political support for safer communities. And the old can lead social movements on behalf of the young, as have a cadre of "grannies" worldwide who are using their free time to advocate policies to avert climate change.

It may seem a tall order for our age to make the shift from recklessness to foresight. The seeds of change, however, lie in the many choices that all of us can make today. We can create cultural practices, institutions, laws, and norms that build concern for future generations. Our individual decisions matter, from how we vote to how we interact in our neighborhoods and communities to how we contribute to our shared culture. As we learn to look further ahead in our own lives and work, we can each become more powerful constituents for reclaiming the future as a society.

HOPE FOR A RECKLESS AGE

We must come to see that the end we seek is
a society at peace with itself, a society that
can live with its conscience.
—MARTIN LUTHER KING, JR.

I can vividly remember warm days in late spring when I was in elementary school. Bugs flickered across the playground puddles, and birds caroused in the forsythia bushes. My classmates and I writhed at our desks, ready to escape for the summer. We anticipated it, welcomed it. I pictured myself climbing into the family station wagon for the ten-hour drive that would deliver us to a friend's beach house, the exotic license plates and highway billboards I'd spy out the rear window, the taste of the salt as I'd plunge myself into the ocean. I made plans: which stuffed animals to pack, which swimsuit to wear, which friends I'd send postcards to, and how I'd persuade my older siblings to let me play their card game.

Even as adults, when we look forward to something in the future that we have experienced before, we can feel it in our bones almost as if we were smelling the fresh bread outside a bakery. We see ourselves in the future making decisions; we feel empathy with ourselves in an imagined state of elation or exuberance. We make decisions, buying a guidebook for a vacation or planning the guest list for a wedding. We imagine the encounters we'll have; we inhabit them.

The ability to engage with the desired future is reinforced by culture and by institutions. Shared experiences with friends and family, advertisements, and media allow us to envision a wedding day as happy; tax credits encourage us to move toward marriage, a commitment that for many people lasts for decades into the future.

When we dread something in the future that is out of our direct control or past experience, on the other hand, we often feel anxiety and even paralysis. It happens when we think about getting older and weaker, when we contemplate an earthquake in our city, and when we think about the loss of our favorite beaches or forests to a changing climate. Our culture and our institutions offer us little collective sense of how to engage with the futures that cause us dread, except as passive recipients of prediction—as victims.

As I finished this book, I came to realize that in order to become less reckless as people and as a society, we must learn how to weigh threats and opportunities of the future with neither paralysis nor naiveté. We need the means to plan for disastrous futures, rather than turn away from them. I had to find a way to think about myself in old age, for example, with empathy rather than aversion. Collectively, we have to be able to plan for the threat of the next Dust Bowl, rather than bury our heads in the dirt. We need to shift from ignoring—or anxiously avoiding—the future in these instances to anticipating it with clear-eyed wisdom.

This is not easy when we constantly get reminded of just how much there is to dread about the future. We see signs of the dangers ahead every day: A refugee crisis is looming. The national debt is growing. The Arctic is melting, and the seas will rise. These are futures nobody wants.

The dire warnings we hear about the future reflect humanity's growing knowledge about the longevity of our actions and the threats that lie around the corner. And yet so often, we don't draw on such scientific knowledge to avert crisis. We have lacked a way to think about unsavory futures with imagination, empathy, and agency. As a result, we seldom turn our knowledge of the future into the power to shape it.

Understanding this has altered how I think about climate change, and about the ongoing task of helping communities and businesses prepare for a hotter, more chaotic future on the planet.

Intelligent people disagree about how to deal with a changing climate—about whether we should take drastic action to stop fossil fuel emissions or to prepare for floods, droughts, and heat waves. The political factions embroiled in this controversy have predominantly offered us three visions of the future: The climate doomsayers tell us that without action, humanity will face a dystopic nightmare of destructive storms, virulent diseases, and crowded, sweltering cities. The advocates of inaction, on the other hand, tell us that stopping emissions will cripple economies and millions will lose their jobs in coal mines and oil fields. The alternative to both of these pictures of the future is, by default, a future that looks much the same as today.

I have been as guilty as anyone of trafficking in doomsday scenarios when talking to business leaders and communities about climate change. They are not entirely useless: Dystopic scenes from science fiction, after all, can at times allow people to imagine horrors we'd prefer not to entertain. Yet when I hear these scenarios for the planet, I certainly don't want to inhabit them. Most people don't tend to linger in visions of the future worse—or even equal to—the present. We don't imagine and empathize with our future selves, or see our communities and societies solving problems, when we dread something. Instead, we feel paralyzed. It's not a coincidence that we don't do much individually or collectively to plan ahead for the futures we imagine under climate change. Either we don't take them seriously, and are overly optimistic, or we find them too overwhelming to even contemplate.

Painting these future scenarios, of course, has served a purpose—conveying the urgency to act on climate change or to prevent action, respectively. The irony is that the politics of climate change has in this way blocked the perception of our individual and collective ability to address the crisis. We're culturally and institutionally reinforcing views of the future that no one can look forward to with a sense of being able to do something.

I now see that we need new ways of looking at the future in light of the changing climate, where we see ourselves taking steps to make our communities, businesses, and societies better in light of real threats. We need to be able to see ourselves making the best out of bad situations and improving the average or good situations. This does not mean sanitizing the dangers posed by climate change, but rather picturing ourselves taking realistic steps to address them. We need scenarios where the future looks better than either now or the past—not simply the same or worse—based on actions we take. Climate role-play games, in which communities make decisions to avert catastrophe are one way to achieve this.

Our cultural practices, media, and institutions must reinforce the future we want. In place of blind denial and paralyzing pessimism, we need a radical strain of optimism. We need to balance urgency with agency, to be armed with tools to confidently look ahead.

There is evidence suggesting that a new kind of optimism can help us address climate change. In a series of eight studies, psychologist Paul Bain and his colleagues asked nearly six hundred Australians to envision the state of society in the year 2050 given a particular scenario. The scenarios asked them to imagine, for example, that severe climate change had been averted, or that marijuana or abortion laws were relaxed. (In Australia, abortion laws vary from state to state.) The study participants—people who practiced different religions, were atheists, or were agnostics—also had varying political persuasions. After writing down details describing the scenario for society they envisioned, they spoke with interviewers about what the collective future would look like. Later, they were asked what policies and personal actions they would support taking today to achieve or avert that possible future.

Bain found that people across each of the studies were most motivated to support current policies or personal behavior change in the immediate future when they imagined a delayed state of increased warmth and morality on the part of people in society—an attribute he called benevolence. This held true whether the people were God-fearing conservatives or progressive atheists.

In a separate study, Bain demonstrated that even climate deniers could be persuaded of the need for "environmental citizenship" if the actions to be taken, such as reducing carbon emissions, were framed as improvements in the way people would treat one another in the imagined future. A collective idea of the future in which people work together on environmental problems, and are more caring and considerate—or a future with greater economic and technological progress—motivated the climate change deniers to support such actions even when they didn't believe that human-caused climate change was a problem.

This research is hardly the final word on the subject, not least because it involved hypothetical as opposed to real-world decisions. Yet it suggests a new way to summon visions of the future, in which neighbors work together and communities thrive while addressing climate change. It also implies that it is not enough to envision technologies like solar panels or wind farms. In order to motivate people to act today, we need to imagine human beings as more compassionate. Like the leaders of successful social movements of the past, political leaders and cultural arbiters might do well to offer the public animating images of neighborhoods where people help the elderly and the young survive heat waves, of cities that have easy-to-use public transit and verdant parks where people gather for common cause. The prophecies of doomsayers might be replaced by realistic scenarios of future climate change that portray humans and communities not as victims, but as agents of change that work together with friends and strangers. I am not suggesting that singing "Kumbaya" around the campfire will solve climate change, but I am suggesting that cultivating a sense of collective agency could motivate more people to make choices today for the sake of the future, whether it's how they vote, eat, use energy, or influence others.

To paint the future in new ways, however, will not be enough to secure a better future. We also need to know how to make decisions in light of uncertain threats and opportunities that lie ahead.

Here, we can draw on the insights of those who have exercised foresight against the odds to point the path forward. The stories and research I unearthed for this book reveal five key lessons for our reckless age. Each is a strategy that has specific tactics enumerated—both to help us see the long road ahead and to stay the course.

1. Look Beyond Near-Term Targets. *We can avoid being distracted by short-term noise and cultivate patience by measuring more than immediate results.* As individuals, we can avoid single data points as measures of our progress or success in life or work, and instead adopt rituals of reflection on long-range goals. Organizations, as well as communities and societies, can use multiple metrics, and look for closer proxies of what they want to ultimately achieve. Within investment firms, corporations, aid organizations, we can ask questions to get at underlying opportunities and threats. All of us can track trends in data over longer spans of time, rather than looking at singular snapshots, to get a sense of how we are faring not just in a moment, but in the long run. Looking up frequently from the minutiae of work and life, and keeping organizations small, can help with this.

2. Stoke the Imagination. *We can boost our ability to envision the range of possibilities that lie ahead.* As individuals, we can create more anchors to the future that allow us to imagine it, whether it's perennial gardens we cultivate, letters we write to our future selves or descendants, or sensory experiences of future risks such as those enabled by virtual reality and other technologies. We can also leave ourselves time and bandwidth to let our minds wander to generate scenarios of the future, and practice if/ then rituals that allow us to imagine ourselves navigating those scenarios with agency. Organizations and communities can cultivate imagination about future risks and opportunity through reverse stress tests and prospective hindsight practices, and through postgame rehashes, in which they bring delayed outcomes to life for people. Prize competitions can be used to activate the imagination of inventors and problem solvers who might not otherwise pursue future-oriented feats. Role-play games, carried out by groups diverse in age, experience, and perspective, can help communities and organizations find opportunities for agency even amid disaster and to inhabit scary scenarios instead of sanitizing them. Businesses and societies can also draw on past examples that cover long spans of history and multiple possibilities to vet potential futures. Concrete visions of the future can animate social movements to overcome obstacles in the present.

3. Create Immediate Rewards for Future Goals. *We can find ways to make what's best for us over time pay off in the present.* As individuals and families, we can reward ourselves for progress toward future goals, or seek programs that offer immediate allure but are designed for our long-run interest, such as plans that link lotteries to building a savings account. Businesses can find ways to apply long-range research to immediate challenges, and invent or use technologies like perennial grains that bear fruit in the near term while sustaining their purpose over time. Innovators and investors can make it more attractive to fund groundbreaking research by combining fast-return and slow-return investments into new portfolios. Communities and societies can use the strategy to offer immediate payoffs to citizens when adopting future-oriented social policies, such as pairing near-run dividends with taxes on pollution.

4. Direct Attention Away from Immediate Urges. *We can reengineer cultural and environmental cues that condition us for urgency and instant gratification.* As individuals, we can seek environments and subcultures where we are not subject to temptation overload, as in the casino. Organizations can use dedicated teams and technologies to interrupt decisions where urgency prevails, such as inappropriately prescribing an antibiotic or cutting corners in programming a rocket. They can also create environments and cultures to cue people to make wiser decisions, including by giving them more slack and setting social norms. Communities and societies can offer more public funding for campaigns and reform their campaign finance laws to reduce the pressure on politicians to act on immediate profit motives. This should be a major focus of people fighting for greater foresight in society. In the meantime, leaders can use delay tactics to make decisions with greater foresight.

5. Demand and Design Better Institutions. *We can create practices, laws, and institutions that foster foresight.* As individuals, we can vote and advocate for rules and policies that encourage looking ahead rather than reckless decisions, such as catch shares that make fishermen more like long-range investors and legal frameworks that protect communities from lawsuits when they exercise precaution about dangerous development. We need analogous rules to encourage investors to hold on to stocks for longer periods, to compensate CEOs for long-run rather than immediate progress, and to reduce the pressure of itinerant shareholders on company leaders and boards. As communities and societies, we need to put our most valued resources into collective stewardship as heirlooms, in the vein of national parks and World Heritage sites, to be held in trust for present and future generations and enforced by law.

While we can take some of these actions today in our personal lives without much trouble, others will require us to exercise our power as investors, voters, business leaders, teachers, consumers, community members, and beyond. Not all of it, of course, will be easy. But none of it is beyond our reach.

What the examples in this book suggest is that at smaller scales, communities and organizations have greater success in exercising foresight. At a time when many institutions are growing in size and scope, when companies merge and our idea of a community can extend across oceans, we may need to rely more heavily on what remains of small communities and groups, and family-owned enterprises, to guide us. Large organizations and societies will need leaders with the courage to define culture and build better environments.

My own source of hope derives from having learned that even in this reckless age, we are not powerless. What I once believed was inevitable about human nature is rather the result of decisions we have made in the past. How we now use our power to shape the future is a choice.

Acknowledgments

I was lucky to work with the brilliant, compassionate, and wise Jake Morrissey at Riverhead on this project. He is the exact editor the book needed—and the kind every writer deserves. (To Jake's daughters: I hate to admit it, but the guy is often right.) I am grateful to Geoff Kloske for championing this project. I am also in awe of the extraordinary works he publishes year after year and of his whip-smart team. Katie Freeman, Kevin Murphy, Jackie Shost, Jynne Martin, Kate Stark, Shailyn Tavella, Lydia Hirt, Mary Stone, and Jessica White are each exceptional human beings, and I am brimming with appreciation for their talents and their guidance. I thank Muriel Jorgensen for her diligent and smart copyediting.

Flip Brophy, literary agent extraordinaire, represented me and this project with integrity and enthusiasm, and supported me when it mattered most. Don't tell my friends this, but she might be my favorite person in New York to meet for lunch.

I was a fellow at New America while working on this book, which was a greatly nourishing experience. I found my intellectual mother ship in its community of thoughtful journalists, policy geeks, and authors. My sincere thanks to Anne-Marie Slaughter, Peter Bergen, Awista Ayub, Fuzz Hogan, and Kirsten Berg—and to Andrés Martinez, Ed Finn, and Torie Bosch for later inviting me to join the cabal of Future Tense.

Eric Lander, my longtime mentor, supported this book in immeasurable ways. I am deeply grateful to him for all of it, and to the Broad Institute. Cornelia Dean, who has been my mentor and friend for more than eleven years, understood this book's importance immediately and has been of great help along the way.

It's safe to say I would have never written this book if not for the brilliant advice and buoying insight of Martha Sherrill and Bill Powers. They helped me at a critical point during the project's birth, and Martha, a savant of books and publishing, guided me through the thorniest points of the process. It is insufficient to simply thank them, but still necessary.

Nothing gives me more hope for the future than experiencing the kindness of strangers. Many, many people took time to share their stories, insights, and life's work with me for this book—even when they barely knew me or did not know me at all. Some of these people's names appear in the preceding pages, but I am grateful to far more than those named for their time and generosity. I want to especially thank the following strangers and friends who spoke to me about their areas of expertise, reviewed parts of the manuscript, and plied me with stories, research, and contacts: Mullen Taylor, Rebecca Darr Litchfield, Alison Loat, Amy Mowl, Dan Honig, Rob Kirschen, Carrie Freeman, Feizal Satchu, Jeri Weiss, Elisabeth Rhyne, Daniel Rozas, Doug Rader, Robert Jones, Pamela Hess, Selam Daniel, Jennifer Shahade, Tim Crews, Julia Szymczak, Greg Flynn, and David Rosenthal.

I greatly appreciate Kristen Zarrelli, who sent me journal articles faster than I could finish describing them, a testament to her unfailing competence and to the value of the library sciences. The astounding and talented John Kenney lent the manuscript his exacting eye. I am also grateful to Elizabeth Shreve for all her efforts.

Marcella Bombardieri read an early version of the book with her characteristic brilliance, skepticism, and precision, while she was moving houses with a toddler, writing a magazine article, and working a full-time job in education policy. Her perceptive reflections and advice vastly improved the book while proving that she is, as suspected, superhuman.

The extraordinary Deborah Blum gave me expert book publishing advice and read my manuscript with boundless generosity and razor-sharp insight. Deborah's formidable intellect is matched only by the magnitude of her kindness.

A decade ago, Benjamin Lambert had no idea that becoming my friend meant condemning himself to later reading my terrible early drafts and listening to my writing woes—he now knows that he was thus cursed. He has been an invaluable intellectual interlocutor and fellow literary traveler from the moment this book was a germ-sized idea. My friend of two decades Mary Bulan continues to serve her life sentence of inspiring me with her music and science, reacting to my work, and reminding me that the world is full of wonder. She also almost killed me on a lake in a lightning storm just before I turned in the manuscript; I want to thank her for not succeeding prematurely.

I was fortunate enough to work alongside the fiercely creative and intelligent Anne Lilly during two writing retreats; her profoundly moving works of visual art and her artistic courage fortified me. She also served as a reader, attuning me to critical fault lines. Lisa Peck helped me see my book from a new perspective, and lent it and me her tremendous compassion, wide-ranging expertise, and insight about cultural and political divides.

Drew Houpt advised me with his exceptional intuition for narrative and his deep understanding of the project and the creative industries. I am lucky to have a close friend who is so talented, generous, and funny, and who is always down for talking books and movies over bourbon.

Several friends gave me homes away from home while I wrote and reported this book. I offer huge and heartfelt thanks to Bill and Barbara Bennett; Mark Bittman and Kathleen Finlay; Peter and Belinda O'Brien; and Sandra Ubuong Saunders.

When I set out to write this book, I harbored illusions about what the next few years would hold. I had no idea how challenging or thrilling they would be, how many conversations and investigations and places would be involved, and how unwieldy it would be to make sense of it all. I am grateful to the people in my life who walked alongside me. Their support has, at times, felt like more than one person deserves. Rachel Relle and Martha Leary consistently reminded me why the project mattered, and rejuvenated me with their warmth and wisdom. My cousin Dhivya Venkataraman imparted her signature wit, smarts, and generosity of spirit. I relied on Manny Gonzales, Julianne Ortega, and Jennifer Galvin for creative banter, smart reactions, and comic relief. I have no words to adequately thank Tiffany Sankary for her creative insights and compassion or Nina Friedman for her profound wisdom and expert guidance over the past few years. I am deeply grateful for the enthusiastic support, superb advice, and levity imparted by: Seth Mnookin, Lori Cole, Nithya Venkataraman, Cyndi Stivers, Miguel Ilzarbe, David Bennett, Marydale DeBor, Xochitl Gonzalez, Amanda Cook, Daria Bishop, Karen Lee Sobol, Bulbul Kaul, Chris Agarwal, Alisha Blechman, Errol Morris, Julia Sheehan, Mary Katherine O'Brien, Kiera Bulan, Tracy Kukkonen, Jeff Perrin, Angela Borges, Adam Grant, Juliette Berg, Ellen Clegg, Jessica Hinchliffe, Gigi Hirsch, Adriana Raudzens, YiPei Chen-Josephson, Brendan Rose, Erin Marotta, Lisa Camardo, Tom Zeller, Ari Ratner, Manette Jungels, Taylor Milsal, Kate Ellis, Charlotte Morgan, Linda Ziemba, Pamela Reeves, Lori Lander, Mary Cleaver, Cheryl Effron, Sarovar Banka, Nicole St. Clair Knobloch, Jeff Goodell, Kristin McArdle, Irene Hamburger, Joel Janowitz, Sunny Bates, Tennessee Grimes, Kim Larson, Yvonne Abraham, Kristina Costa, Jen LaCroix, and Bill Fish.

This book is dedicated to my parents, for many reasons. For instance, they perennially believe I can do things, plus they had the decency to pretend they didn't care when I quit law school. My grandmother Seetha Rajagopal has given me intangible heirlooms that I treasure as much as her father's dilruba. I also thank my wonderful siblings. My sister Dini Rao marked milestones with great enthusiasm, ideas, and bottles of wine. My brother Avi Garg cared deeply about the project, riffing with me till the wee hours and reading an early draft with an eagle eye. My brother-in-law Anju Rao responded to half-baked title options with consistent candor and humor. My nieces Neela and Ayoni virtually got me out of bed to write (and occasionally, literally woke me up) on the darkest days of 2016 and 2017—it was their future I kept imagining.

I am beyond grateful to my partner, Andrew Fish, for inspiring me with his creative discipline, putting up with my thousand-yard stares, and reminding me to eat over the past few years. Not everyone gets to take for granted someone so spectacular. His presence, unwavering support, and astounding works of art have made my life far better than I could have expected.

Notes

CHAPTER 1. GHOSTS OF THE PAST AND FUTURE

15 **The commander of the Roman naval fleet:** I relied primarily on two sources to reconstruct the events of 79 CE: Mary Beard, *The Fires of Vesuvius: Pompeii Lost and Found* (Cambridge, MA: Belknap Press of Harvard University Press, 2010); and Pliny the Younger, *The Letters of the Younger Pliny*, trans. Betty Radice (Harmondsworth, England: Penguin Books, 1969).

16 **In the year 62 CE . . . Seneca:** Lucius Annaeus Seneca, *Naturales Quaestiones*, trans. Thomas H. Corcoran (Cambridge, MA: Harvard University Press, 1971). Some scholars question whether the date of the earthquake was 62 or 63.

17 **The ancient Romans routinely searched for signs:** Susanne William Rasmussen, *Public Portents in Republican Rome* (Rome: "L'Erma" di Bretschneider, 2003).

17 **the oft-bemoaned weather forecast:** Nate Silver, *The Signal and the Noise: The Art and Science of Prediction* (New York: Penguin, 2012), esp. pp. 125–28.

19 **A 2006 survey:** For the survey, see Bruce Tonn, Angela Hemrick, and Frederick G. Conrad, "Cognitive Representations of the Future: Survey Results," *Futures* 38, no. 7 (2006): 810–29. In addition, a 2017 survey of more than twenty-eight hundred American adults conducted by the Institute for the Future showed that more than half never thought about the future beyond thirty years, and that more than a quarter either rarely or never think about their future five years ahead. Age did not make them more or less likely to think about the future. See "The American Future Gap," Institute for the Future, 2017, http://www.iftf.org/fileadmin/user_upload/downloads/IFTF_TheAmerican FutureGap_Survey_SR-1948.pdf.

19 **Humans have roamed the Earth:** See Gregg Easterbrook, "What Happens When We All Live to 100?" *The Atlantic*, October 2014; and Trevor Stokes, "Human Life Span Took Huge Jump in Past Century," LiveScience, October 15, 2012, https://www.livescience.com/23989-human-life-span -jump-century.html.

20 **Hal Hershfield:** I interviewed Hershfield on September 15, 2015, and tried the Aging Booth app shortly afterward. The study described can be found in Hal E. Hershfield et al., "Increasing Saving Behavior Through Age-Progressed Renderings of the Future Self," *Journal of Marketing Research* 48, no. SPL (November 2011), https://doi.org/10.1509/jmkr.48.spl.s23.

   In a separate study of young adults, Hershfield and colleagues found that people who wrote letters to their future selves or encountered virtual reality versions of themselves were less likely to make delinquent choices, such as buying stolen goods or cheating on a task: Jean-Louis Van Gelder, Hal E. Hershfield, and Loran F. Nordgren, "Vividness of the Future Self Predicts Delinquency," *Psychological Science* 24, no. 6 (2013): 974–80, https://doi.org/10.1177/0956797612465197.

22 **Today, meteorologists can map out:** See Silver, *The Signal and the Noise*, and Erik Larson, *Isaac's Storm: A Man, a Time, and the Deadliest Hurricane in History* (New York: Crown, 1999).

22 **How people react to severe hurricane forecasts:** Robert J. Meyer and Howard Kunreuther, *The Ostrich Paradox: Why We Underprepare for Disasters* (Philadelphia: Wharton Digital Press, 2017). See also Howard Kunreuther, Robert Meyer, and Erwann Michel-Kerjan, "Overcoming Decision Biases to Reduce Losses from Natural Catastrophes," 2009, http://opim.wharton.upenn.edu/risk

/library/C2009_HK,RJM,EMK.pdf. Meyer and Kunreuther point out the difference in forecasting technology available between the 1900 hurricane that took Galveston, Texas, by surprise and killed more than eight thousand people—the deadliest natural disaster in U.S. history—and Hurricane Ike, which hit the region in 2008. Between the two storms were major advances in hurricane prediction, and a seventeen-foot seawall had been constructed to protect the city. Yet even with warnings before Ike hit the Texas Gulf Coast, most of the city's homes and businesses were damaged, at a cost of $14 billion, and a hundred people died. Less than 40 percent of the thousands of homeowners whose houses were damaged by floodwaters had purchased flood insurance in advance. The seawall, viewed as an eyesore to would-be coastal dwellers, had, ironically, encouraged building booms north and south of the city along the coast. These areas away from the seawall sustained even greater losses. Thousands of residents lost their homes.

22 **The global cost of hurricane damage:** This is assuming ongoing climate change. See Robert Mendelsohn et al., "The Impact of Climate Change on Global Tropical Cyclone Damage," *Nature Climate Change* 2, no. 3 (2012): 205–9, https://doi.org/10.1038/nclimate1357.

23 **In the United States, only 10 percent:** Howard Kunreuther and Erwann Michel-Kerjan, "Natural Disasters," in *Global Problems, Smart Solutions: Costs and Benefits,* ed. Bjørn Lomborg (Cambridge, England, and New York: Cambridge University Press, 2013), 439–65.

23 **From 1960 to 2011, more than 60 percent:** Helga Fehr-Duda and Ernst Fehr, "Sustainability: Game Human Nature," *Nature* 530, no. 7591 (2016): 413–15, https://doi.org/10.1038/530413a.

23 **Dennis Mileti:** This conversation took place on April 13, 2017.

23 **Jay Segarra:** The quotations and the account derive from my interview with Segarra on November 16, 2017. He is the father of a friend and former classmate of mine. I also consulted a blog where Segarra recounted his experience during Hurricane Katrina; see *The Katrina Experience: An Oral History Project,* http://www.thekatrinaexperience.org/?p=22.

25 **People often take in information:** Robert J. Meyer et al., "The Dynamics of Hurricane Risk Perception: Real-Time Evidence from the 2012 Atlantic Hurricane Season," *Bulletin of the American Meteorological Society* 95, no. 9 (2014): 1389–1404, https://doi.org/10.1175/bams-d-12-00218.1. See also Meyer and Kunreuther's *The Ostrich Paradox,* cited above, for more on the optimism bias and on insurance buying rates pre- and post-disaster.

26 **At airports in the 1950s and 1960s:** Helga Fehr-Duda and Ernst Fehr, "Sustainability: Game Human Nature," *Nature* 530, no. 7591 (2016): 413–15, https://doi.org/10.1038/530413a.

26 **"a monopoly of the visible":** Wim Wenders, "The Visible and the Invisible," lecture presented at "Wide Angle: The Norton Lectures on Cinema," April 2, 2018.

26 **"availability bias":** I relied primarily on Daniel Kahneman's discussion of this concept and its foundational research, especially in chapter 13 of his *Thinking, Fast and Slow* (New York: Farrar, Straus and Giroux, 2011). In addition to the availability bias, Kahneman describes research findings that demonstrate how people perceive immediate threats as more likely than cumulative risks over time, which contributes to inadequately heeding long-run and future consequences.

People also misunderstand predictions when the action to be taken is inconvenient in the short run. Studies of volcano warnings show that people given a forecast that says an eruption is likely within a two-week span often mistakenly believe that the odds of the eruption are higher after thirteen days than on day one of the time span, and they will delay plans to evacuate as a result.

Some features of modern prediction might actually be making things worse when it comes to imagining the future. One of the hallmarks of today's advanced scientific predictions is that they are not offered as foolproof prophecies of the future, but as estimates of how likely or unlikely certain events might be, over given time frames. In other words, what sets scientific predictions apart from the divinations of oracle bone readers and octopuses is the open disclosure of how little is known about the future. Weather reports tell us there is an 80 percent chance of rain tomorrow—not that it will definitely rain. Forecasters show us the predicted path of a hurricane by portraying a "cone of uncertainty," meant to convey to the public that no one knows for sure where the storm will go, but the most likely paths are contained within the shaded region. Research conducted by Bob Meyer, surveying thousands of coastal residents as they watched forecasts for approaching hurricanes, has shown that people take more action to prepare for storms when they see a dotted line of the most likely hurricane path—as opposed to a cone of uncertainty. This happens even if people live far afield from the specified path. It seems people act when given specific predictions they can latch on to—even if less accurate than a range of possible scenarios.

28 **mental time travel:** I interviewed Thomas Suddendorf on this topic on July 25, 2017. I interviewed Daniel Schacter on August 3, 2017. I consulted extensive literature on this topic, among other

sources: Martin E. P. Seligman et al., *Homo Prospectus* (Oxford: Oxford University Press, 2017); Daniel L. Schacter et al., "The Future of Memory: Remembering, Imagining, and the Brain," *Neuron* 76, no. 4 (2012): 677–94, https://doi.org/10.1016/j.neuron.2012.11.001; Thomas Suddendorf and Michael C. Corballis, "The Evolution of Foresight: What Is Mental Time Travel, and Is It Unique to Humans?" *Behavioral and Brain Sciences* 30, no. 3 (2007): 299–313, https://doi.org /10.1017/s0140525x07001975; Daniel L. Schacter, Roland G. Benoit, and Karl K. Szpunar, "Episodic Future Thinking: Mechanisms and Functions," *Current Opinion in Behavioral Sciences* 17 (2017): 41–50, https://doi.org/10.1016/j.cobeha.2017.06.002; Adam Bulley, Julie D. Henry, and Thomas Suddendorf, "Thinking About Threats: Memory and Prospection in Human Threat Management," *Consciousness and Cognition* 49 (2017): 53–69, https://doi.org/10.1016/j.concog.2017.01.005; Pascal Boyer, "Evolutionary Economics of Mental Time Travel?" *Trends in Cognitive Sciences* 12, no. 6 (2008): 219–24, https://doi.org/10.1016/j.tics.2008.03.003; Gabriele Oettingen, A. Timur Se-vincer, and Peter M. Gollwitzer, *The Psychology of Thinking About the Future* (New York: Guilford Press, 2018).

29 **scenario generation and mind-wandering:** I interviewed Tracy Gleason on June 7, 2016. The study referred to on mind-wandering is Benjamin Baird, Jonathan Smallwood, and Jonathan W. Schooler, "Back to the Future: Autobiographical Planning and the Functionality of Mind-Wandering," *Consciousness and Cognition* 20, no. 4 (2011): 1604–11, https://doi.org/10.1016/j.concog.2011.08.007.

30 **when you invite people to generate detailed scenes:** This concept encapsulates a range of research studies, most from experimental settings, and is informed by my interviews of the psychologists noted above. Relevant studies include: Adam Bulley, Julie Henry, and Thomas Suddendorf, "Prospection and the Present Moment: The Role of Episodic Foresight in Intertemporal Choices Between Immediate and Delayed Rewards," *Review of General Psychology* 20, no. 1 (2016): 29–47, https://doi.org/10.1037/gpr0000061; Uli Bromberg, Antonius Wiehler, and Jan Peters, "Episodic Future Thinking Is Related to Impulsive Decision Making in Healthy Adolescents," *Child Development* 86, no. 5 (2015): 1458–68, https://doi.org/10.1111/cdev.12390; Sara O'Donnell, Tinuke Oluyomi Daniel, and Leonard H. Epstein, "Does Goal Relevant Episodic Future Thinking Amplify the Effect on Delay Discounting?" *Consciousness and Cognition* 51 (2017): 10–16, https://doi.org/10.1016 /j.concog.2017.02.014; Sarah E. Snider, Stephen M. LaConte, and Warren K. Bickel, "Episodic Future Thinking: Expansion of the Temporal Window in Individuals with Alcohol Dependence," *Alcoholism: Clinical and Experimental Research* 40, no. 7 (2016): 1558–66, https://doi.org/10.1111 /acer.13112; Fania C. M. Dassen et al., "Focus on the Future: Episodic Future Thinking Reduces Discount Rate and Snacking," *Appetite* 96 (2016): 327–32, https://doi.org/10.1016/j.appet.2015.09.032; Yan Yan Sze et al., "Bleak Present, Bright Future: Online Episodic Future Thinking, Scarcity, Delay Discounting, and Food Demand," *Clinical Psychological Science* 5, no. 4 (2017): 683–97, https://doi.org/10.1177/2167702617696511.

31 **Social movements:** I interviewed Marshall Ganz on several occasions in early 2017. Nichelle Nichols recounts the story of meeting Martin Luther King, Jr., in "A Conversation with Nichelle Nichols," StarTalk Radio, 2017, https://www.startalkradio.net/show/a-conversation-with-nichelle-nichols.

32 **a group of retired NASA scientists:** Carolyn Y. Johnson, "Elder Scientists Work to Send Humans to Mars." *The Boston Globe*, February 8, 2015, https://www.bostonglobe.com/metro/2015/02/08/elder -statesmen-science-unite-for-mars-mission/N5sZQqOEuhKC56rdtE4uPN/story.html.

33 **Stanford's Virtual Human Interaction Lab:** I visited and interviewed researchers at the lab on April 7, 2016. I also consulted: Jeremy Bailenson, "Virtual Reality Could Make Real Difference in Envi-ronment," *SFGate*, August 15, 2014, https://www.sfgate.com/opinion/article/Virtual-reality-could -make-real-difference-in-5691610.php; and Jeremy Bailenson, "Infinite Reality: The Dawn of the Virtual Revolution," Stanford+Connects lecture, October 9, 2013, https://www.youtube.com /watch?v=1jbwxR8bCb4.

36 **Anab Jain:** See Anab Jain, "Why We Need to Imagine Different Futures," TED Talk, April 2017, https://www.ted.com/talks/anab_jain_why_we_need_to_imagine_different_futures.

36 **wiser decisions with respect to the future:** See, for example, Lisa Zaval, Ezra M. Markowitz, and Elke U. Weber, "How Will I Be Remembered? Conserving the Environment for the Sake of One's Legacy," *Psychological Science* 26, no. 2 (2015): 231–36, https://doi.org/10.1177/0956797614561266.

36 **Researchers in Germany similarly found:** Sabine Pahl and Judith Bauer, "Overcoming the Dis-tance," *Environment and Behavior* 45, no. 2 (2011): 155–69, https://doi.org/10.1177/0013916511417618.

37 **"Death Over Dinner":** I met Michael Hebb through a mutual friend. Details on the effort are dis-cussed in Richard Harris, "Discussing Death Over Dinner," *The Atlantic*, April 18, 2016, https:// www.theatlantic.com/health/archive/2016/04/discussing-death-over-dinner/478452/.

38 **Danny Hillis:** I interviewed Danny Hillis and attended Long Now talks where the clock was described in 2016 and 2017.

40 **In Halberstadt, Germany:** Daniel J. Wakin, "John Cage's Long Music Composition in Germany Changes a Note," *The New York Times*, May 6, 2006, https://www.nytimes.com/2006/05/06/arts/music /06chor.html.

40 **the Voyager capsules:** "What Are the Contents of the Golden Record?" NASA, accessed August 13, 2018, https://voyager.jpl.nasa.gov/golden-record/whats-on-the-record/.

CHAPTER 2. DASHBOARD DRIVING

43 **By 2020, we'll have tens of billions:** This figure has been widely cited as fifty billion, but it's recently been disputed by some observers as an overestimate. The source for the fifty billion figure is Dale Evans, "The Internet of Things: How the Next Evolution of the Internet Is Changing Everything," Cisco Systems, 2011, https://www.cisco.com/c/dam/en_us/about/ac79/docs/innov/IoT _IBSG_0411FINAL.pdf.

43 **Dan Falk:** See especially Dan Falk, *In Search of Time: The History, Physics and Philosophy of Time* (New York: Thomas Dunne Books/St. Martin's Griffin, 2008), esp. p. 80.

44 **Vijay Mahajan and the history of Indian microfinance:** My telling of Mahajan's story stems from interviews with him and other microfinance experts from India and abroad, and from a detailed journal and blog he kept during his pilgrimage. I also read interviews and watched video footage of his speeches and statements both before and after the crisis. I am indebted to many experts who helped me understand the history and underlying causes of the 2010 Indian microfinance crisis, including Bindu Ananth, Amy Mowl, Elisabeth Rhyne, Veena Mankar, Daniel Rozas, and others who preferred not to speak on the record. The deep reporting on the crisis by journalist Ketaki Gokhale at *The Wall Street Journal*, cited more specifically below, was also of great help. I consulted the following sources, among many others, to fill in my research on Mahajan and BASIX: Arvind Ashta, "Dealing with Black Swan Events: An Interview with Vijay Mahajan, Founder and CEO of Basix," *Strategic Change* 25, no. 5 (2016): 625–39, https://doi.org/10.1002/jsc.2085; Cyril Fouillet and Britta Augsburg, "Profit Empowerment: The Microfinance Institution's Mission Drift," *Perspectives on Global Development and Technology* 9, no. 3–4 (2010): 327–55, https://doi.org/10.1163 /156914910X499732; Prabhu Ghate, "Consumer Protection in Indian Microfinance: Lessons from Andhra Pradesh and the Microfinance Bill," *Economic and Political Weekly* 42, no. 13 (2007): 1176– 84, http://www.jstor.org/stable/4419417; Philip Mader, "Rise and Fall of Microfinance in India: The Andhra Pradesh Crisis in Perspective," *Strategic Change* 22, no. 1–2 (2013): 47–66, https://doi.org /10.1002/jsc.1921; Vijay Mahajan and T. Navin, "Microfinance in India: Growth, Crisis and the Future," BASIX report, 2012; M. S. Sriram, "Commercialisation of Microfinance in India: A Discussion of the Emperor's Apparel," *Economic and Political Weekly* 45, no. 24 (2010): 65–73.

48 **Muhammad Yunus:** For more detail, see Connie Bruck, "Millions for Millions," *The New Yorker*, October 30, 2006, https://www.newyorker.com/magazine/2006/10/30/millions-for-millions.

48 **he turned BASIX into a holding company:** Many other microfinance institutions were making similar transitions at the same time. Also, it's worth noting that in between founding BASIX and his prior work, Mahajan did work with self-help groups via an organization called Pradan.

49 **In the 1980s, aided by nonprofit organizations:** For this history, I consulted Vijay Mahajan and T. Navin, "Microfinance in India: Lessons from the Andhra Crisis," chapter 1 of Doris Köhn, *Microfinance 3.0: Reconciling Sustainability with Social Outreach and Responsible Delivery* (Heidelberg: Springer, 2013).

50 **More than $500 million:** See Mader, "Rise and Fall of Microfinance in India."

50 **High repayment rates on loans:** The large for-profit lenders were seeing 99 percent repayment rates, whereas the sector as a whole was seeing 97 percent repayment rates.

50 **In 2008 and 2009, the ten largest microfinance companies:** See Mader, "Rise and Fall of Microfinance in India."

50 **In 2009, the state was home:** N. Srinivasan, *Microfinance India: State of the Sector Report 2009* (New Delhi: SAGE Publications India, 2009).

51 **SKS Microfinance Limited:** For details on SKS and Vikram Akula, I relied on interviews, with the sources on the Indian microfinance crisis noted above, and Eric Bellman and Arlene Chang, "India's Major Crisis in Microlending," *The Wall Street Journal*, October 28, 2010, https://www.wsj.com /articles/SB10001424052702304316404575580663294846100; Mader, "Rise and Fall of Microfinance in India"; and Shloka Nath, "At the Crossroads," *Forbes*, September 27, 2009, https://www

.forbes.com/2009/09/25/crossroads-vikram-akula-sks-microfinance-suresh-gurumani-forbes-india
.html#22b031ed39b6.

51 **Akula is a controversial figure:** This was corroborated in numerous conversations with experts in
microfinance within India and abroad. See also the sources cited in the previous note and Vikas Bajaj,
"Amid Scandal, Chairman of Troubled Lender Will Quit," *The New York Times*, November 23, 2011,
https://www.nytimes.com/2011/11/24/business/global/vikram-akula-chairman-of-sks-microfinance
-to-step-down.html.

51 **"This work can be driven only by greed":** Eric Bellman, "Entrepreneur Gets Big Banks to Back Very
Small Loans," *The Wall Street Journal*, May 15, 2006, https://www.wsj.com/articles/SB1147654
89678552599.

52 **Yunus worried:** On Yunus's criticism of Akula and his model, see Megha Bahree, "Microfinance or
Loan Sharks? Grameen Bank and SKS Fight It Out," *Forbes*, September 21, 2010, https://www
.forbes.com/sites/meghabahree/2010/09/21/microfinance-or-loan-sharks-grameen-bank-and-sks
-fight-it-out/#3cf7148a54cc; and Nath, "At the Crossroads."

52 **All eyes in the microfinance field:** The source for this paragraph is an interview with Beth Rhyne,
December 2015.

53 **"polar vortex":** See Caitlyn Kennedy, "Wobbly Polar Vortex Triggers Extreme Cold Air Outbreak,"
National Oceanic and Atmospheric Administration, January 8, 2014, https://www.climate.gov/news
-features/event-tracker/wobbly-polar-vortex-triggers-extreme-cold-air-outbreak; and Ian Living-
ston, "Polar Vortex Delivering D.C.'s Coldest Day in Decades, and We're Not Alone," *The Washington Post*, January 7, 2014, https://www.washingtonpost.com/news/capital-weather-gang/wp/2014/01
/07/polar-vortex-delivering-d-c-s-coldest-day-in-decades-and-were-not-alone/?noredirect=on&
utm_term=.cc9223b34534.

53 **Temporary dips in gas prices:** Brad Tuttle, "Gas Prices Are Falling at an Incredible Rate—and Why
That's a Problem," *Time*, January 14, 2016, http://time.com/money/4180721/gas-prices-decline
-electric-cars/; and Chris Isidore, "Low Gas Prices Boost SUV and Pickup Sales," CNNMoney,
December 4, 2015, accessed August 13, 2018, https://money.cnn.com/2015/12/04/autos/gas-prices
-suv-pickup-sales/index.html.

53 **Studies of New York City cabdrivers:** There have been conflicting results from studies of New York
taxi drivers on this point. Nevertheless, the balance of research supports the idea that cabdrivers
adopt an informal daily earnings target. Two initial studies, one from New York and one from Singapore, support my conclusion: C. Camerer et al., "Labor Supply of New York City Cabdrivers: One
Day at a Time," *The Quarterly Journal of Economics* 112, no. 2 (1997): 407–41, https://doi.org/10.1162
/003355397555244. The most persuasive counterinterpretation of cabdrivers' habits comes from
Henry Farber at the National Bureau of Economic Research, who published a working paper indi-
cating that cabdrivers may quit when they are tired each day, not when they fulfill a particular target. See Henry S. Farber, "Is Tomorrow Another Day? The Labor Supply of New York City
Cabdrivers," *Journal of Political Economy* 113, no. 1 (2005): 46–82, https://doi.org/10.1086/426040.

However, an empirical study with a larger and more recent data set that included a review of
the previous research reconciled the results and convinced me that the practice of using an informal
daily target is likely at play; see Ender Faruk Morgul and Kaan Ozbay, "Revisiting Labor Supply of
New York City Taxi Drivers: Empirical Evidence from Large-Scale Taxi Data," Transportation
Research Board, 2014, available at http://engineering.nyu.edu/citysmart/trbpaper/15-3331.pdf. Another study using Farber's data supports the argument that cabdrivers adopt an informal daily earnings target; see Vincent P. Crawford and Juanjuan Meng, "New York City Cab Drivers' Labor
Supply Revisited: Reference-Dependent Preferences with Rational-Expectations Targets for
Hours and Income," *American Economic Review* 101, no. 5 (2011): 1912–32, https://doi.org/10.1257
/aer.101.5.1912.

A *New York Times* story published in 2017 revealed that the ride-hailing companies Uber and
Lyft had seized upon the target-setting habit of drivers to try to keep more cars in circulation at
high demand times, so they could meet customers' requests for rides more quickly. Uber had used
video game and slot machine techniques, including a graphical gauge that shows drivers falling
short of an arbitrary earnings target they have not set themselves, and pop-up reminders that con-
tinually update drivers about how much farther they have to go to reach the target or what they
stand to lose if they quit working for the day. They found these strategies can change decisions
drivers make, keeping them working longer hours than they would otherwise.

The conclusion I draw is further supported by the widespread evidence that people manage
their efforts toward short-term targets across many contexts.

The fear of missing a target, for instance, also dampens the performance of professional golfers—people we would expect to do everything they can to get the best scores. In a Wharton Business School study of golf tournaments on the PGA tour that tracked more than 2.5 million putts, researchers found that pros will desperately try to avoid scoring a bogey (one stroke above par) when putting at a given hole, even if it means sacrificing the opportunity for a birdie (one stroke below par). Since a golfer's score depends only on the overall number of strokes in the tournament across seventy-two holes, it makes sense to put equal focus on every putt—whether or not it is for par. But that doesn't shake the feeling that hitting above par at a given hole feels like a loss to golfers—they are locked on that near-term target when they are in front of it, and not thinking of the overall tournament. When the pros putt for birdies—which are just icing on the cake and not to avoid feeling loss—they seem to put less focus on doing well. Professional putts for birdies are hence less accurate than putts to avoid bogeys. This pattern of putting more effort into the shots to avoid bogeys than to get birdies does not pay off in the end—it costs players on average one stroke per tournament game. It turns out that even the best pro golfers do this. Tiger Woods openly admitted in 2007 that it was more important to him to make putts for par than birdies psychologically, although not in terms of his actual score. (This point is further elaborated in chapter 5 and its endnotes.)

54 **aversion to losing:** On prospect theory and loss aversion, see Daniel Kahneman, *Thinking, Fast and Slow* (New York: Farrar, Straus and Giroux, 2011).

55 **credit crisis in the making:** Ketaki Gokhale, "A Global Surge in Tiny Loans Spurs Credit Bubble in a Slum," *The Wall Street Journal*, August 14, 2009, https://www.wsj.com/articles/SB12501211251 8027581.

The warning signs were not limited to rumors, media reports, or even expert commentary. They were emerging signals from the field, in India, that showed that borrowers and local government officials did not have the same perception as the most vocal industry leaders.

In 2006, in the Krishna district of Andhra Pradesh, accounts of overindebted poor people being harassed by microfinance lenders and shamed by their fellow self-help group members surfaced. People who were struggling to repay loans from multiple companies complained of being coerced by loan officers and forced to flee their villages. Reports of suicide emerged, following incidents in which borrowers in default had notices posted, or group meetings convened, outside their front doors. That year, a local district official closed fifty-seven branches of two major microfinance companies, and borrowers were told not to repay their loans. The incident was limited in geographic scope to one district, but still had a significant impact on at least two lenders. The companies faced consequences from policies of zero tolerance for less-than-perfect loan repayment rates.

Other countries had already seen credit bubbles in microfinance burst on the eve of the Indian crisis. Incidents in Morocco and Bolivia pointed to how precarious microlending could be if borrowers took on too much debt. In 2009, a crisis emerged in Nicaragua, where rapid loan growth fueled by foreign investment had led to borrowers taking on too many loans. The No Pago (I won't pay) movement, to stop paying back microloans, rose up and was backed by the Nicaraguan president.

55 **In a series of articles:** See in particular the following articles by Ketaki Gokhale: "As Microfinance Grows in India, So Do Its Rivals," *The Wall Street Journal*, December 15, 2009, http://online.wsj .com/article/SB126055117322287513.html; and "Group Borrowing Leads to Pressure," *The Wall Street Journal*, August 13, 2009, http://online.wsj.com/article/SB125008232217325553.html.

55 **Daniel Rozas:** I interviewed Rozas and consulted his commentaries in blogs and journals before and after the crisis. He particularly called out the crisis in the making in "Is There a Microfinance Bubble in South India?" Microfinance Focus, November 17, 2009, http:/ /www.microfinancefocus.com /news/2009/ 11/17/opinion-microfinance-bubble-south-india.

57 **Vikram Akula, on the other hand:** Akula's public position on the eve of the crisis is related in these two pieces he wrote in August 2009: "What Microfinance Crisis in India?" (letter to the editor), *The Wall Street Journal*, August 19, 2009, https://www.wsj.com/articles/SB1000142405297020468320457 4358321561989110; and "Why There's No Credit Crisis in Microfinance," *Harvard Business Review*, August 25, 2009, https://hbr.org/2009/08/why-theres-no-credit-crisis-in.

57 **The crash came on like a rogue wave:** It's worth noting that the actions of the microfinance lenders and their investors, and their myopia, were not the sole reason that the crash was so spectacular and catastrophic. The Andhra Pradesh government's reaction, to cast out the private lenders and allow borrowers to default on their loans, played a major role. There are several microfinance industry experts who speculate that the state government was more motivated by the rise of private competitors than by the suicides.

58 **"As long as the music is playing"**: Prince's words are quoted in "Ex-Citi CEO Defends 'Dancing' Quote to U.S. Panel," Reuters, April 8, 2010, https://www.reuters.com/article/financial-crisis -dancing-idUSN0819810820100408.

58 **The measure hid the impending danger**: A passage from the Mader analysis, cited above, sums up the situation: "Investments in Indian microfinance enjoyed a reputation as being virtually risk-free thanks to the sector's widely touted loan repayment rate of over 99 percent." It's also worth noting that the microlenders were not measuring factors that might have allowed them to gauge the true risk.

58 **The ancient Greek historian Herodotus wrote about the wise statesman Solon**: While it is an interesting story with resonance, Debra Hamel points out that Herodotus's account of Solon's travels might not be historically accurate. For more detail, see Debra Hamel, *Reading Herodotus: A Guided Tour through the Wild Boars, Dancing Suitors, and Crazy Tyrants of* The History (Baltimore: Johns Hopkins University Press, 2012).

59 **Aristotle echoed the wisdom of Solon**: In chapter X of Book 1 of the *Nicomachean Ethics*, Aristotle refers to the story of Solon. In other writings on ethics, he echoes the point that the measure of a life ought to come at its end.

59 **Thomas Mann**: I read the 1995 Woods translation of Mann's classic for this interpretation of the thermometer readings and of the Director's reaction to Castorp's possible departure. Thomas Mann, *The Magic Mountain*, trans. John E. Woods (New York: Alfred A. Knopf, 1995).

60 **Anne Dias**: This conversation with Dias took place in 2015.

60 **thousands of investors hold on to stocks**: See these articles by Brad M. Barber and Terrance Odean: "The Common Stock Investment Performance of Individual Investors," *SSRN Electronic Journal*, 1998, https://doi.org/10.2139/ssrn.94140; and "Trading Is Hazardous to Your Wealth: The Common Stock Investment Performance of Individual Investors," *The Journal of Finance* 55, no. 2 (2000): 773–806, https://doi.org/10.2139/ssrn.219228.

62 **Mahajan asked former borrowers**: It's worth noting that many of the villagers had other sources of debt beyond the microfinance loans, and used the loans to pay off informal debts to pawnbrokers or other lenders, and vice versa.

64 **Mahajan's rise and fall**: Mahajan has since bounced back and is running companies that are doing well.

CHAPTER 3. BEYOND THE HERE AND NOW

65 **the infamous "marshmallow test"**: For my descriptions of the original experiments and their popular appeal, I relied on Walter Mischel, *The Marshmallow Test: Why Self-Control Is the Engine of Success* (New York: Little, Brown, 2014), and on a variety of media sources.

66 **The studies that have followed**: A variety of studies have complicated the popular conclusions drawn from Mischel's initial experiments. Referred to here in particular is Celeste Kidd, Holly Palmeri, and Richard N. Aslin, "Rational Snacking: Young Children's Decision-Making on the Marshmallow Task Is Moderated by Beliefs About Environmental Reliability," *Cognition* 126, no. 1 (2013): 109–14, https://doi.org/10.1016/j.cognition.2012.08.004.

67 **Bettina Lamm**: For this discussion, I consulted the following: Bettina Lamm et al., "Waiting for the Second Treat: Developing Culture-Specific Modes of Self-Regulation," *Child Development* 89, no. 3 (2018): e261–e277, https://doi.org/10.1111/cdev.12847; Michaeleen Doucleff, "Want to Teach Your Kids Self-Control? Ask a Cameroonian Farmer," *Morning Edition*, NPR, July 3, 2017, https://www .npr.org/sections/goatsandsoda/2017/07/03/534743719/want-to-teach-your-kids-self-control-ask-a -cameroonian-farmer.

68 **Yet another marshmallow study**: Sabine Doebel and Yuko Munakata, "Group Influences on Engaging Self-Control: Children Delay Gratification and Value It More When Their In-Group Delays and Their Out-Group Doesn't," *Psychological Science* 29, no. 5 (2018): 738–48, https://doi.org /10.1177/0956797617747367.

69 **Timothy Flacke**: The quotations and description of Flacke's work come from interviews I conducted with him in 2017.

69 **When the Federal Reserve surveyed**: This percentage is the one that motivated Flacke, which is why I cite the 2015 data. In 2016, the figure dropped to 44 percent and in 2017, to 41 percent. See Board of Governors of the Federal Reserve System, "Report on the Economic Well-Being of U.S. Households in 2017," May 2018, https://www.federalreserve.gov/publications/files/2017-report -economic-well-being-us-households-201805.pdf.

69 **People in desperation:** For this discussion I consulted S. Mullainathan and E. Shafir, *Scarcity: The New Science of Having Less and How It Defines Our Lives* (New York: Picador Books, 2014); the quotation is from p. 109.

70 **Bindu Ananth:** I interviewed Ananth on three occasions from 2015 to 2017 about the Indian microfinance crisis and related reforms. Ananth told me the loan repayment rate is today still the predominant gauge of whether microfinance programs in India are succeeding in helping the poor and whether the lending companies are thriving. Yet as recently as 2016, her organization, Dvara Trust (formerly IFMR Trust) published a study showing that underlying near-perfect loan repayment rates among the poor are high numbers of households in financial distress, who still cannot afford the number of loans they have been given. Many poor borrowers will do whatever it takes to repay their loans to formal lenders to keep from being stigmatized by delinquency. In the process, they take on informal debt from neighbors, pawn jewelry, or stretch their meager family budgets too thin. When a child gets sick or a bill needs to be paid, families can fall into inescapable cycles of debt. Nevertheless, in 2018, she wrote to tell me: "There is a new industry effort [that is] trying to measure financial health/financial well-being as an outcome rather than product metrics such as number of loans disbursed. I am hopeful that even though this is harder to measure and fuzzier, it will serve as a better metric of our collective progress."

71 **In 2009, the Michigan Credit Union League:** For more detail, see the excellent reporting in Rob Walker, "How to Trick People into Saving Money," *The Atlantic*, May 2017, https://www.theatlantic.com/magazine/archive/2017/05/how-to-trick-people-into-saving-money/521421/.

72 **pilot in Michigan:** For this discussion, I interviewed Peter Tufano and consulted the following: Shawn Allen Cole, Benjamin Charles Iverson, and Peter Tufano, "Can Gambling Increase Savings? Empirical Evidence on Prize-Linked Savings Accounts," August 8, 2017, Saïd Business School WP 2014-10, available at SSRN: https://ssrn.com/abstract=2441286 or http://dx.doi.org/10.2139/ssrn.2441286; Mauro F. Guillén and Adrian E. Tschoegl, "Banking on Gambling: Banks Lottery-Linked Deposit Accounts," *Journal of Financial Services Research* 21. no. 3 (2002): 219–31, https://doi.org/10.1023/A:1015081427038; Sebastian Lobe and Alexander Hölzl, "Why Are British Premium Bonds So Successful? The Effect of Saving with a Thrill," *SSRN Electronic Journal*, 2007, https://doi.org/10.2139/ssrn.992794; Peter Tufano, "Saving Whilst Gambling: An Empirical Analysis of UK Premium Bonds," *American Economic Review* 98, no. 2 (2008): 321–26, https://doi.org/10.1257/aer.98.2.321; Peter Tufano, Jan-Emmanuel De Neve, and Nick Maynard, "U.S. Consumer Demand for Prize-Linked Savings: New Evidence on a New Product," *Economics Letters* 111, no. 2 (2011): 116–18, https://doi.org/10.1016/j.econlet.2011.01.019.

74 **glitter approach:** On the topic of short-term rewards helping with long-term goals, see, for example, Kaitlin Woolley and Ayelet Fishbach, "Immediate Rewards Predict Adherence to Long-Term Goals," *Personality and Social Psychology Bulletin* 43, no. 2 (2016): 151–62, https://doi.org/10.1177/0146167216676480.

74 **One expert on flood readiness:** The expert referred to is Lisa Sharrard, described in greater detail in chapter 7. I interviewed her in 2017.

75 **Until 1990, people in California who made home improvements:** See Robert J. Meyer and Howard Kunreuther, *The Ostrich Paradox: Why We Underprepare for Disasters* (Philadelphia: Wharton Digital Press, 2017), p. 100.

75 **The glitter approach might even offer:** There is a reason the United States has not adopted such strategies to fight climate change, and it has to do in part with the fact that no one proposed them quickly enough. From the initial years when scientists sounded the alarm about global climate change, a concerted effort on the part of fossil fuel companies and their proxies has taken place to characterize climate change in turns as not really happening, as not caused by humans, or as a problem that would ruin the global economy to solve. That message got out of the gate long before ideas like giving people carbon dividends came to the fore and became a dominant and misleading meme. Meanwhile, climate scientists were still muddling over how to communicate their uncertainty about the precise rate of change. Some environmentalists who believe the economy must shrink to fight threats to the planet reinforced the message of the fossil fuel advocates that there must be pain and loss in the near term in order to tackle climate change for our grandchildren.

What's needed for climate policy to succeed today are attractive enough glitter explosions not just for individual people but for the powerful constituencies who influence politics. In the United States, especially, we likely need to offer short-term rewards to special-interest groups and members of Congress whose constituents might hold the policy back—whether it's cuts in capital gains or income taxes that counterbalance the higher price of carbon.

   With awareness of what can persuade people to make immediate sacrifices for the long-term good, those of us concerned about climate change might now take greater care to craft proposals that offer everyday people and business owners something good in the short run, rather than just asking them to endure near-term hardship. Although the reality of global climate change is now mired in manufactured controversy, it's not too late to bring people around to a solution that delivers glittery rewards rather than only penalties. It may be an uphill climb, but ultimately worth it.

77 **Casinos take advantage:** For this discussion, I consulted the excellent research of anthropologist Natasha Schüll, and in particular her book *Addiction by Design: Machine Gambling in Las Vegas* (Princeton, NJ: Princeton University Press, 2014).

78 **Like salty bar snacks:** There is an emerging literature comprising many studies on the connection between addiction to devices and increased discounting of delayed consequences. The Swiss psychologist Joel Billieux, for instance, has researched the habits of people who highly depend on their cell phones. In one study of a hundred female college students at the University of Geneva, he found that women who felt most addicted to their mobile devices reported less control of their impulses than those who felt freer of their phones. The phone-dependent students were more likely to decide to make themselves feel better in a moment of impulse that they later regretted. And they were far less prone to be able to stop doing things that later made them feel worse about themselves, like binge-eating or drinking. Students who sent text messages more frequently and spent more time on the phone had more trouble finishing what they started and persisting through tedious tasks toward their long-term goals, such as homework assignments. (While these findings relied on self-reporting in questionnaires, they are consistent with international data on patterns bred by frequent use of digital devices.) My thoughts on this topic are also informed by experience, observation, and this wonderful book: Harold Schweizer, *On Waiting* (London: Routledge, 2008).

78 **Ed Finn:** Ed Finn, "How Predictions Can Change the Future," *Slate*, September 7, 2017, http://www.slate.com/articles/technology/future_tense/2017/09/when_computer_predictions_end_up_shaping_the_future.html.

79 **In the seventeenth century:** Schweizer, *On Waiting*.

79 **We act so much like addicts:** Tony Dokoupil, "Is the Internet Making Us Crazy? What the New Research Says," *Newsweek*, July 9, 2012, https://www.newsweek.com/internet-making-us-crazy-what-new-research-says-65593. See also Adam Alter, *Irresistible: The Rise of Addictive Technology and the Business of Keeping Us Hooked* (New York: Penguin Press, 2017).

80 **professional poker players:** I conducted interviews with about a dozen professional poker players in 2015 and 2016. I am also indebted to Rob Kirschen at the World Series of Poker, who helped me track down useful archival information about tournaments and players.

81 **Ronnie Bardah:** I interviewed Bardah on several occasions; his poker earnings are published by the World Series of Poker.

82 **A pair of British interviewers:** This sample size was too small, in my view, to view this as a scientific finding, but the interview content was interesting and, in many cases, echoed my interviewees' statements. It is published research, nonetheless: Abby McCormack and Mark D. Griffiths, "What Differentiates Professional Poker Players from Recreational Poker Players? A Qualitative Interview Study," *International Journal of Mental Health and Addiction* 10, no. 2 (2011): 243–57, https://doi.org/10.1007/s11469-011-9312-y.

82 **a critical factor is how movement leaders characterize setbacks:** This section derives from interviews with Marshall Ganz and from Karen Beckwith, "Narratives of Defeat: Explaining the Effects of Loss in Social Movements," *The Journal of Politics* 77, no. 1 (2015): 2–13, https://doi.org/10.1086/678531.

83 **no free top-shelf liquor:** This policy can vary, depending on the casino and the state, and free top-shelf liquor can be available to certain elite players. Nonetheless, a poker expert from the World Series of Poker staff told me that most players with elite status "don't take advantage of the higher-quality liquor available to them. They prefer not to drink while playing. Instead, they tend to use this privilege to order Fiji water bottles (as opposed to the ordinary water bottles available to anyone)."

83 **William Powers:** William Powers, *Hamlet's BlackBerry: Building a Good Life in the Digital Age* (New York: Harper Perennial, 2011).

83 **Pico Iyer:** Pico Iyer, "Why We Need to Slow Down Our Lives," Ideas.TED.com, November 4, 2014, https://ideas.ted.com/why-we-need-a-secular-sabbath/.

83 **scheduling an "untouchable day":** Neil Pasricha, "Why You Need an Untouchable Day Every Week," *Harvard Business Review*, March 16, 2018, https://hbr.org/2018/03/why-you-need-an-untouchable -day-every-week.

84 **write letters to sick children, or take time on a Saturday:** Cassie Mogilner, Zoë Chance, and Michael I. Norton, "Giving Time Gives You Time," *Psychological Science* 23, no. 10 (2012): 1233–38, https:// doi.org/10.1177/0956797612442551.

84 **Culture makes a difference:** There is also research suggesting that people in cultures that have similar ways of expressing present and future tense may weigh future consequences and opportunities more heavily than those with sharper divides.

86 **Matt Matros:** I interviewed Matros on several occasions; his poker earnings are published by the World Series of Poker.

87 **Peter Gollwitzer and if/then tactics (implementation intentions):** I interviewed Gollwitzer in 2017 and consulted nearly a dozen studies on implementation intentions, also known as if/then tactics. The following sources give overviews of the research: Peter M. Gollwitzer, "Implementation Intentions: Strong Effects of Simple Plans," *The American Psychologist* 54, no. 7 (1999): 493–503, https:// doi.org/10.1037/0003-066X.54.7.493; Peter M. Gollwitzer, "Weakness of the Will: Is a Quick Fix Possible?" *Motivation and Emotion* 38, no. 3 (2014): 305–22, https://doi.org/10.1007/s11031-014 -9416-3; Gabriele Oettingen, *Rethinking Positive Thinking: Inside the New Science of Motivation* (New York: Current, 2015).

89 **the Shooter Task study:** Saaid A. Mendoza, Peter M. Gollwitzer, and David M. Amodio, "Reducing the Expression of Implicit Stereotypes: Reflexive Control Through Implementation Intentions," *Personality and Social Psychology Bulletin* 36, no. 4 (2010): 512–23, https://doi.org/10.1177/01461672 10362789.

90 **In 2018, the U.S. Government Accountability Office:** U.S. Government Accountability Office, *K-12 Education: Discipline Disparities for Black Students, Boys, and Students with Disabilities. Report to Congressional Requesters*, 2018, GAO-18-258.

90 **When public school teachers in the United States discipline:** I consulted the following research studies on this point, which make clear that the disproportionate discipline is not due to worse behavior on the part of students of particular races: Kent McIntosh et al., "Education Not Incarceration: A Conceptual Model for Reducing Racial and Ethnic Disproportionality in School Discipline," *Journal of Applied Research on Children* 5, no. 2 (2014): Article 4; Russell J. Skiba et al., "The Color of Discipline: Sources of Racial and Gender Disproportionality in School Punishment," *The Urban Review* 34, no. 4 (2002): 317–42, https://doi.org/10.1023/A:1021320817372; Russell J. Skiba et al., "Race Is Not Neutral: A National Investigation of African American and Latino Disproportionality in School Discipline," *School Psychology Review* 40, no. 1 (2011): 85–107.

91 **teachers and "vulnerable decision points":** I interviewed several teachers and school principals, including Ellwood, and I interviewed McIntosh to inform this portion of the chapter. I also watched footage of McIntosh's training workshops for educators and read excerpts of Ellwood's doctoral dissertation, including data documenting the change she saw in her school. See also the specific studies cited above.

CHAPTER 4. THE QUICK FIX

95 **The work of John Graham:** John R. Graham, Campbell R. Harvey, and Shiva Rajgopal, "Value Destruction and Financial Reporting Decisions," *Financial Analysts Journal* 62, no. 6 (2006): 27–39, https://doi.org/10.2469/faj.v62.n6.4351.

96 **Dr. Sara Cosgrove and antibiotic resistance:** The account I give here of Sara Cosgrove's work stems from interviews conducted with her and her colleagues from 2016 to 2018, including a visit to the Johns Hopkins Hospital to observe her and other members of her team on rounds and in meetings. My account of her work and of the general problem of antibiotic prescribing and resistance was informed by interviews with numerous other experts in the field, most notably Neil Fishman and Julia Szymczak.

97 **Superbugs are on the rise:** Review on Microbial Resistance, *Tackling Drug-Resistant Infections Globally: Final Report and Recommendations*, 2016, https://amr-review.org/sites/default/files/160525_Final paper_with cover.pdf.

97 **Contagious tuberculosis:** Here I'm referring to extensively drug-resistant tuberculosis, which at this writing has infected people in 123 countries, and to multidrug-resistant tuberculosis, which kills

hundreds of thousands of people each year. See the WHO's article on tuberculosis at http://www
.who.int/tb/areas-of-work/drug-resistant-tb/global-situation/en/.

98 **The Cocoanut Grove fire and penicillin:** My account of the fire and its contribution to the story of
penicillin was informed by the 1942 report on the fire by the City of Boston's fire commissioner,
William Arthur Reilly; survivor accounts archived by the Cocoanut Grove Coalition at http://www
.cocoanutgrovefire.org; and Stuart B. Levy, *The Antibiotic Paradox: How the Misuse of Antibiotics
Destroys Their Curative Powers* (Reading, MA: Perseus, 2002). I also consulted media accounts of the
fire and of its commemoration from multiple sources. The cause of the fire is still debated.

99 **In the early twentieth century:** President's Council of Advisors on Science and Technology
(PCAST), "Report to the President on Combating Antibiotic Resistance," Executive Office of
the President, 2014, https://www.cdc.gov/drugresistance/pdf/report-to-the-president-on-combating
-antibiotic-resistance.pdf.

99 **In 1927, an Austrian doctor:** See the article on Julius Wagner-Jauregg at https://www.nobelprize
.org/nobel_prizes/medicine/laureates/1927/wagner-jauregg-facts.html.

99 **They also made sex safer:** See Levy, *The Antibiotic Paradox.*

100 **Today a majority of the antibiotics prescribed:** See the PCAST report, cited above. If you expand
this statement to include respiratory tract infections such as sinusitis, strep throat, and pneumonia,
almost half of the antibiotics are not the right treatments, compared with almost all for coughs and
colds.

100 **Alexander Fleming:** C. L. Ventola, "The Antibiotic Resistance Crisis. Part 1: Causes and Threats,"
Pharmacy and Therapeutics 40, no. 4 (2015): 277–83.

101 **Fear drives hospital doctors:** My descriptions of the culture of prescribing were generously informed
by interviews of University of Pennsylvania professor Julia Szymczak and her research comprising
extensive interviews of physicians. See in particular J. E. Szymczak and J. Newland, "The Social
Determinants of Antimicrobial Prescribing: Implications for Antimicrobial Stewardship," in *Practical Implementation of an Antimicrobial Stewardship Program,* ed. T. F. Barlam et al. (Cambridge,
England: Cambridge University Press, 2018).

103 **In a study of more than twenty thousand visits:** Jeffrey A. Linder et al., "Time of Day and the Decision to Prescribe Antibiotics," *JAMA Internal Medicine* 174, no. 12 (2014): 2029, https://doi.org
/10.1001/jamainternmed.2014.5225.

103 **scarcity of time is like scarcity of money:** See S. Mullainathan and E. Shafir, *Scarcity: The New Science
of Having Less and How It Defines Our Lives* (New York: Picador Books, 2014).

104 **"firefighting," via Roger Bohn and Ramachandran Jaikumar:** The findings are summarized in
Roger Bohn, "Stop Fighting Fires," *Harvard Business Review,* July–August 2000, https://hbr.org
/2000/07/stop-fighting-fires. I found the original paper here: https://tomonleadership.com/wp
-content/uploads/firefighting-by-knowledge-workers.pdf.

105 **Ben Taub Hospital and prior approval:** This incident and research are documented in A. Clinton
White, Jr., et al., "Effects of Requiring Prior Authorization for Selected Antimicrobials: Expendi-
tures, Susceptibilities, and Clinical Outcomes," *Clinical Infectious Diseases* 25, no. 2 (1997): 230–39,
http://www.jstor.org/stable/4481112.

106 **Johns Hopkins Hospital and prior approval:** At the hospital, once the decision is made to add an antibiotic to the formulary, the antibiotic stewardship team works with other clinicians in the hospital to
develop recommendations for when and how to use the drugs. The decision to stock a drug on-site at
the hospital depends on what kinds of infections patients are appearing with, what superbugs are
emerging with resistance to which drugs, and how effective and affordable the drugs are for patients.

107 **Most antibiotics are prescribed:** PCAST, "Report to the President on Combating Antibiotic
Resistance."

108 **Daniella Meeker:** The three interventions described are documented in the two studies cited below,
which are among a broader literature that evaluates such behavioral approaches. There is also research
demonstrating that the effects of at least two of these interventions endured past the study period.
Daniella Meeker et al., "Effect of Behavioral Interventions on Inappropriate Antibiotic Prescribing
Among Primary Care Practices," *JAMA* 315, no. 6 (2016): 562, https://doi.org/10.1001/jama.2016.0275;
and Daniella Meeker et al., "Nudging Guideline-Concordant Antibiotic Prescribing," *JAMA Internal
Medicine* 174, no. 3 (2014): 425, https://doi.org/10.1001/jamainternmed.2013.14191.

110 **In 2014, the British government:** Michael Hallsworth et al., "Provision of Social Norm Feedback to
High Prescribers of Antibiotics in General Practice: A Pragmatic National Randomised Controlled
Trial," *The Lancet* 387, no. 10029 (2016): 1743–52, https://doi.org/10.1016/s0140-6736(16)00215-4.

110 **California has since tried:** April Dembosky, "New California Database to Help Doctors Spot Prescription Drug Abuse," KQED Science, December 30, 2015, https://www.kqed.org/stateofhealth /132232/new-california-database-to-help-doctors-spot-prescription-drug-abuse.

111 **Benjamin Franklin:** The descriptions of events in Franklin's life come largely from his public writings, and particularly his autobiography. I consulted the version edited by John Bigelow for this purpose.

111 **Upon his death in 1790:** Fox Butterfield, "From Ben Franklin, a Gift That's Worth Two Fights," *The New York Times*, April 21, 1990, https://www.nytimes.com/1990/04/21/us/from-ben-franklin-a-gift -that-s-worth-two-fights.html.

113 **Franklin and daylight saving time:** Benjamin Franklin, letter to the editor, *The Journal of Paris*, 1784, accessed at http://www.webexhibits.org/daylightsaving/franklin3.html. Also found in *The Ingenious Dr. Franklin. Selected Scientific Letters*, ed. Nathan G. Goodman (Philadelphia: University of Pennsylvania Press, 1931), pp. 17–22. This idea was not only a joke to Franklin, as some have suggested. The same notion had occurred to him while strolling the Strand and Fleet Street in London at seven in the morning, where he saw the shuttered shops of merchants who complained of taxes on candles, and wondered why they did not keep the hours of the sun. He was committed enough to his proposal to make sure he got credit for it for posterity, putting it into his autobiography.

113 **The original purpose of observing daylight saving time has since been rendered moot:** Matt Schiavenza, "Time to Kill Daylight Saving," *The Atlantic*, March 8, 2015, https://www.theatlantic.com /national/archive/2015/03/time-to-kill-daylight-saving/387175/.

114 **Maria Montessori:** Two biographies of Montessori's life, complemented by interviews of experts in her educational method, informed my description of her. The better of the biographies was E. M. Standing, *Maria Montessori: Her Life and Work* (London: Hollis & Carter, 1957). This more hagiographic biography filled in certain gaps: Rita Kramer, *Maria Montessori: A Biography* (New York: Da Capo Press, 1988).

115 **Hal Gregersen and Montessori's results:** Gregersen initially described his result in this interview: Bronwyn Fryer, "How Do Innovators Think?" *Harvard Business Review*, September 28, 2009, https://hbr.org/2009/09/how-do-innovators-think. See also Jeffrey Dyer, Hal Gregersen, and Clayton M. Christensen, *The Innovator's DNA: Mastering the Five Skills of Disruptive Innovators* (Boston: Harvard Business Review Press, 2011).

Studies of Montessori *students* are limited in scope and size. But one interesting study published in the journal *Science* in 2006 compared three-year-old children from low-income families in Milwaukee, selected by lottery for public Montessori preschool, with children from similar backgrounds sent to more conventional public preschools. At age five and at age twelve, the Montessori students exhibited greater social problem-solving skills and less proclivity for conflict, greater creativity, and equal or better performance on standardized tests as students who went to other schools. A marked difference appeared at age five in Montessori students: higher executive functioning, the skill that aids people with impulse control and with valuing the future in decisions. In other words, foresight. The finding is not definitive proof, in my view, but enough reason to take a close look at the method.

116 **what happened at the bus stops:** I also once witnessed this scheme at play at a post office in a rural village in Italy. And a version of it has been adopted by Southwest Airlines, which gives its passengers designated places to stand in line, rather than jockeying for positions, while waiting to board its planes.

117 **Richard Larson and waiting in lines:** Ana Swanson, "What Really Drives You Crazy About Waiting in Line (It Actually Isn't the Wait at All)," *The Washington Post*, November 27, 2015, https://www .washingtonpost.com/news/wonk/wp/2015/11/27/what-you-hate-about-waiting-in-line-isnt-the -wait-at-all/?postshare=2761449015328760&utm_term=.032ca55984a0.

118 **"nudges":** Richard H. Thaler and Cass R. Sunstein, *Nudge: Improving Decisions About Health, Wealth and Happiness* (New York: Penguin, 2009).

119 **Tristan Harris:** I found information about his work via his website: http://www.tristanharris.com/.

119 **a similar approach was taken up in the Black Forest:** D. Pichert and K. V. Katsikopoulos, "Green Defaults: Information Presentation and Pro-Environmental Behaviour," *Journal of Environmental Psychology* 28, no. 1 (2008): 63–73, https://doi.org/10.1016/j.jenvp.2007.09.004.

Another way to use choice architecture for climate change policy would be to commit in advance to future increases in fossil fuel prices or cuts to energy use—to ratchet up as the economy or gross domestic product of the country grows. This is like the workplace schemes in which people commit to increase the amount they are saving prospectively—whenever they get a raise at work. This anchors a sacrifice to an overall increase of resources, so that it no longer seems like a loss of something. It also shifts the sacrifice away from the current moment of temptation to a more distant time horizon, when research shows people are more willing to make sacrifices.

The residents of Zurich, Switzerland, showed in 2008 that it can be politically appealing to make this kind of policy, when they voted by a large margin in support of becoming a "2,000-watt society." Zurich's citizens agreed to restrict the city's energy use over time, so that it would drop to 1960s levels by 2050. The concrete, shared goal at the relatively intimate scale of a city became a source of pride. Economists Helga Fehr-Duda and Ernst Fehr point out that by 2016, Zurich residents were using less than half the amount of energy used by most people in the United States. The effort's success hinged in part on precommitting to later decreases in energy use—delaying the pain of cutting back. See Helga Fehr-Duda and Ernst Fehr, "Sustainability: Game Human Nature," *Nature* 530, no. 7591 (2016): 413–15, https://doi.org/10.1038/530413a.

119 **creating "slack":** See Mullainathan and Shafir, *Scarcity.*

120 **"temporary problem solvers":** See Bohn, "Stop Fighting Fires."

120 **a kind of postgame rehash:** Several studies indicate the effectiveness of this method: E. Avdic et al., "Impact of an Antimicrobial Stewardship Intervention on Shortening the Duration of Therapy for Community-Acquired Pneumonia," *Clinical Infectious Diseases* 54, no. 11 (2012):1581–87; D. X. Li et al., "Sustained Impact of an Antibiotic Stewardship Intervention for Community-Acquired Pneumonia," *Infection Control & Hospital Epidemiology* 37, no. 10 (2016): 1243–46; P. D. Tamma et al., "What Is the More Effective Antibiotic Stewardship Intervention: Preprescription Authorization or Postprescription Review with Feedback?" *Clinical Infectious Diseases* 64 (2017): 537–43, http://dx .doi.org/10.1093/cid/ciw780. A similar program has been adopted with success at the University of Pennsylvania's teaching hospital, as I learned in an interview with Dr. Neil Fishman, its stewardship director and infectious disease expert.

CHAPTER 5. A BIRD'S-EYE VIEW

124 **Eagle Capital and the Currys' story:** I constructed this story and timeline on the basis of extensive interviews with the team at Eagle Capital and conversations with their clients. I also consulted the firm's internal documents to corroborate figures shared with me.

  Author's disclosure: Ravenel and Beth Curry's son Boykin, described later in the chapter, sits on the board of New America, a nonprofit organization that gave me a fellowship in 2015. I met Boykin nearly a year after I was awarded the fellowship. He had no influence over the decision to award me a fellowship, nor did anyone at New America suggest that I write about Eagle Capital or the Currys. For that matter, no one on New America's staff or board has had editorial input or control over the contents of this book.

126 **Peter Rothschild:** I interviewed Rothschild in April 2017 to corroborate the story told to me by Ravenel Curry.

127 **Goodhart's law:** A good source on this topic is Jerry Z. Muller, *The Tyranny of Metrics* (Princeton, NJ: Princeton University Press, 2018). See also Dan Honig, *Navigation by Judgment: Why and When Top Down Management of Foreign Aid Doesn't Work* (New York: Oxford University Press, 2018).

128 **In the nineteenth century, Chinese peasants:** S. Mullainathan and E. Shafir, *Scarcity: The New Science of Having Less and How It Defines Our Lives* (New York: Picador Books, 2014).

128 **David Deming and student test scores:** David J. Deming et al., "When Does Accountability Work? Texas System Had Mixed Effects on College Graduation Rates and Future Earnings," *Education Next* 16, no. 1 (2016): 71–76. See also Liz Mineo, "School Testing a Mixed Bag, Study Says," *The Harvard Gazette*, October 30, 2015, https://news.harvard.edu/gazette/story/2015/10/school-testing -a-mixed-bag-study-says/.

129 **The corporate practice of projecting next quarter's profits:** For this discussion, I consulted a range of reports and studies and conducted interviews with dozens of investors, corporate board leaders, academics, and corporate executives. On the topics of sacrificing future goals for the quarter, forgoing R&D, increased pressure for short-term results, and share buybacks, see the following, which include data for described survey results: Jonathan Bailey and Jonathan Godsall, "Short-Termism: Insights from Business Leaders: Findings from a Global Survey of Business Leaders Commissioned by McKinsey & Company and CPP Investment Board," CPPIB and McKinsey & Company, 2013, http://www.shareholderforum.com/access/Library/20131226_McKinsey.pdf; Dominic Barton, "Capitalism for the Long Term," *Harvard Business Review*, March 2011, https://hbr.org/2011 /03/capitalism-for-the-long-term; Dominic Barton, Jonathan Bailey, and Joshua Zoffer, "Rising to the Challenge of Short-Termism," FCLT Global, 2016, https://www.fcltglobal.org/docs/default -source/default-document-library/fclt-global-rising-to-the-challenge.pdf; Joseph L. Bower and Lynn S. Paine, "The Error at the Heart of Corporate Leadership," *Harvard Business Review*, May

25, 2017, https://hbr.org/2017/05/managing-for-the-long-term; Rebecca Darr and Tim Koller, "How to Build an Alliance Against Corporate Short-Termism," McKinsey & Company, 2017, https://www.mckinsey.com/business-functions/strategy-and-corporate-finance/our-insights/how-to-build-an-alliance-against-corporate-short-termism; Alex Edmans, "Study: When CEOs' Equity Is About to Vest, They Cut Investment to Boost the Stock Price," *Harvard Business Review*, February 28, 2018, https://hbr.org/2018/02/study-when-ceos-equity-is-about-to-vest-they-cut-investment-to-boost-the-stock-price; John R. Graham, Campbell A. Harvey, and Shiva Rajgopal, "The Economic Implications of Corporate Financial Reporting," *Journal of Accounting and Economics* 40 (2005): 3–73; John R. Graham, Campbell R. Harvey, and Shiva Rajgopal, "Value Destruction and Financial Reporting Decisions," *Financial Analysts Journal* 62, no. 6 (2006): 27–39, https://doi.org/10.2469/faj.v62.n6.4351; William Lazonick, "Profits Without Prosperity," *Harvard Business Review*, September 2014, https://hbr.org/2014/09/profits-without-prosperity; Robert N. Palter, Werner Rehm, and Jonathan Shih, "Communicating with the Right Investors," *McKinsey Quarterly*, April 2008, https://www.mckinsey.com/business-functions/strategy-and-corporate-finance/our-insights/communicating-with-the-right-investors.

131 **Yet we are harmed in less sensational:** See Graham, Harvey, and Rajgopal, "Value Destruction and Financial Reporting Decisions."

131 **Focusing Capital on the Long Term and Sarah Keohane Williamson:** My accounts are based on interviews with Williamson and her team, and events I attended. The study described of 615 corporations is Dominic Barton et al., "Measuring the Economic Impact of Short-Termism," McKinsey Global Institute, 2017, https://www.mckinsey.com/~/media/mckinsey/featured%20insights/long%20term%20capitalism/where%20companies%20with%20a%20long%20term%20view%20outperform%20their%20peers/mgi-measuring-the-economic-impact-of-short-termism.ashx. The study also showed that cumulative earnings of the long-term-oriented companies grew on average by 36 percent more than earnings of the short-term-oriented companies.

   Note: While I have carefully considered the views of skeptics who dispute that short-term thinking in corporate America is a problem, data like these prove far more persuasive than their insistence that as long as at least some American companies are still reaping high profits, we should bury our heads and hope for the best like the inhabitants of Easter Island on the verge of their own demise.

132 **Eric Ries:** Eric Ries, "Vanity Metrics vs. Actionable Metrics—Guest Post by Eric Ries," the blog of author Tim Ferriss, May 19, 2009, https://tim.blog/2009/05/19/vanity-metrics-vs-actionable-metrics/.

134 **A company like Amazon:** I recommend this analysis by an Andreessen Horowitz venture capital investor: Benedict Evans, "Why Amazon Has No Profits (and Why It Works)," Andreessen Horowitz, May 22, 2018, https://a16z.com/2014/09/05/why-amazon-has-no-profits-and-why-it-works/.

134 **Warren Buffett famously ignores short-term noise for the truly long term:** See Buffett's February 2018 annual letter to shareholders, where he reflects on the lessons learned from the bet: http://www.berkshirehathaway.com/letters/2017ltr.pdf. The bet was originally $1 million, but because both sides eventually invested their stakes in Berkshire Hathaway stock, the donation to the charity exceeded $2 million.

134 **Seth Klarman:** The price of Klarman's investing book seems to vary over time, with estimates from $795 for a used copy to more than $3,000 for a new copy. See Brad Tuttle, "Meet the Billionaire Investing Guru Whose 27-Year-Old Book Is Now Selling for $3,000" *Time*, July 16, 2018, http://time.com/money/5338595/seth-klarman-margin-of-safety-author-value-investing/; and Andrew Ross Sorkin, "A Quiet Giant of Investing Weighs In on Trump," *The New York Times*, February 7, 2017, https://www.nytimes.com/2017/02/06/business/dealbook/sorkin-seth-klarman-trump-investors.html.

136 **Google X and "pre-mortems":** Astro Teller, "The Head of 'X' Explains How to Make Audacity the Path of Least Resistance," *Wired*, April 15, 2016, https://www.wired.com/2016/04/the-head-of-x-explains-how-to-make-audacity-the-path-of-least-resistance/.

136 **Deborah Mitchell and "prospective hindsight":** Deborah J. Mitchell, J. Edward Russo, and Nancy Pennington, "Back to the Future: Temporal Perspective in the Explanation of Events," *Journal of Behavioral Decision Making* 2, no. 1 (1989): 25–38, https://doi.org/10.1002/bdm.3960020103. See also Gary Klein, "Performing a Project Premortem," *Harvard Business Review*, September 2007, https://hbr.org/2007/09/performing-a-project-premortem.

139 **An Ernst & Young survey showed:** These data were analyzed in Dominic Barton and Mark Wiseman, "Focusing Capital on the Long Term," *Harvard Business Review*, January–February 2014, https://hbr.org/2014/01/focusing-capital-on-the-long-term.

141 **Dan Honig:** I interviewed Honig in 2018. See also his book *Navigation by Judgment*, cited above.

142 **Paul Polman and Unilever:** Unilever was headquartered in both London and Rotterdam when Polman took over, and has since consolidated the headquarters in the Netherlands. I interviewed Jeff Seabright, chief sustainability officer at Unilever, in April 2017. For more on Paul Polman and Unilever, see Matthew Boyle, "Unilever Hands CEO Polman $722,000 Bonus for Sustainability Work," Bloomberg.com, March 7, 2014, https://www.bloomberg.com/news/articles/2014-03-07/unilever -hands-ceo-polman-722-000-bonus-for-sustainability-work; and Michael Skapinker, "Can Unilever's Paul Polman Change the Way We Do Business?" *Financial Times*, September 29, 2016, https://www .ft.com/content/e6696b4a-8505-11e6-8897-2359a58ac7a5. In spring 2018, Polman began facing shareholder dissent over executive compensation, unrelated to the sustainability question.

142 **Royal DSM:** Hugh Welsh, "An Insider's View: Why More Companies Should Tie Bonuses to Sustainability," *The Guardian*, August 11, 2014, https://www.theguardian.com/sustainable-business /2014/aug/11/executive-compensation-bonuses-sustainability-goals-energy-water-carbon-dsm.

143 **Jeff Bezos:** I read Bezos's 1997 shareholder letter on the U.S. Securities and Exchange Commission website (https://www.sec.gov/Archives/edgar/data/1018724/000119312516530910/d168744dex991 .htm), though it is widely available through a variety of sources. See also Evans, "Why Amazon Has No Profits (and Why It Works)," cited above.

144 **Investors, for their part, could also choose alternative metrics:** Dominic Barton, a former global managing partner of McKinsey & Company, suggests looking at ten-year economic value added, R&D efficiency, numbers of patents in the pipeline, and multiyear return on capital investments. See Barton and Wiseman, "Focusing Capital on the Long Term," where Barton raises these ideas.

146 **Demis Hassabis:** I met Hassabis in 2015, when I first asked him about AI, and I also interviewed him by phone in 2016.

147 **Goldman Sachs has already automated:** Kevin Maney, "Goldman Sacked: How Artificial Intelligence Will Transform Wall Street," *Newsweek*, February 26, 2017, https://www.newsweek.com /2017/03/10/how-artificial-intelligence-transform-wall-street-560637.html.

147 **Howard Marks:** Howard Marks, *The Most Important Thing: Uncommon Sense for the Thoughtful Investor* (New York: Columbia Business School Publishing, 2011).

148 **A series of automated decisions:** The risks of high-frequency trading conducted by computers at prominent firms have become more apparent in recent years. Artificial intelligence expert and author David Auerbach writes: "In 2012, the Knight Capital Group deployed a high-frequency trading algorithm to make automated stock trades at high speed, faster than humans could oversee. Because of a bug, the system did not register its trades as complete, so it kept making them. In less than an hour, the $300 million firm was bankrupt. A month later, Peet's Coffee soared 5 percent in the two minutes after the Nasdaq's opening bell, causing the Nasdaq to cancel many of the trades during that interval; the culprit was never revealed. The 2010 'flash crash' was triggered by similarly unregulated high-frequency trade behavior by fund trader Waddell and Reed, except there it snowballed to encompass the whole market, forcing the exchanges to undo thousands and thousands of transactions. HFT was blamed for a similar 'flash crash' on February 5, 2018, when the Dow dropped 800 points in ten minutes because of automated responses to an increase in derivative volatility." Auerbach, personal correspondence.

149 **Buddy Guindon and the history of the Galveston fishery:** My account was informed by interviews with Buddy Guindon and other Gulf fishermen who generously shared their time with me, as well as interviews with fishery experts and scientists, including Doug Rand and Robert Jones of the Environmental Defense Fund, to whom I am grateful. It was also informed by data from the National Oceanic and Atmospheric Administration and research studies on fishery management policy.

150 **the "tragedy of the commons" and Elinor Ostrom:** Garrett Hardin, "The Tragedy of the Commons," *Science* 162, no. 3859 (1968): 1243–48, https://doi.org/10.1126/science.162.3859.1243, http://science .sciencemag.org/content/162/3859/1243; Elinor Ostrom, *Governing the Commons: The Evolution of Institutions for Collective Action* (Cambridge, England: Cambridge University Press, 1990).

151 **tragedies of the commons might also be viewed from the lens of time horizons:** Daniel Altman of NYU's Stern School of Business and Jonathan Berman of J.E. Berman Associates have made a related point about what is commonly known as corporate social responsibility. These are efforts that companies make to improve communities where they work or to conserve natural resources. Many companies that adopt philanthropic "do-gooder" efforts often relegate them to low-level managers and give them small budgets and short shrift, instead of seeing them as core to the business's ability to keep earning profits over time. Altman and Berman believe that a longer time horizon would correct for the deeper problem, which is that companies don't see a near-term return from making

such investments. A company that invests in a community sanitation program when it builds a road, they argue, will eventually be more likely to secure a contract to build another road and should incorporate such long-run benefits of social engagement in its business decisions. Altman and Berman posit that companies like Nike or Tata group make significant and sustained investments in community initiatives because they know they will see eventual profits from doing so. Of course, there are plenty of exceptions—companies that have longtime horizons but have bought up the water rights of a region for themselves or waged campaigns to deny the science of climate change. A long perspective offers some potential, but it's no guarantee that a business will do what's right for society. See Daniel Altman and Jonathan Berman, "The Single Bottom Line," 2011, http://library.businessethicsworkshop.com/images/Library/sblfinal%20corporate%20citizen%20and%20profit.pdf; and Daniel Altman and Jonathan Berman, "Explaining the Long-Term Single Bottom Line (SSIR)," *Stanford Social Innovation Review*, June 24, 2011, https://ssir.org/articles/entry/explaining_the_long-term_single_bottom_line.

152 **The catch share scheme for Gulf Coast red snapper:** The comeback of the fishery was documented by the federal government's fisheries program overseen by the National Oceanic and Atmospheric Administration, confirmed by the Environmental Defense Fund's science and policy experts, and corroborated by the fishermen I interviewed. Nevertheless, the success of the commercial catch share scheme is now endangered by the excess of recreational fishing for red snapper, which has no such system in place.

152 **A study published in the journal *Nature*:** Anna M. Birkenbach, David J. Kaczan, and Martin D. Smith, "Catch Shares Slow the Race to Fish," *Nature* 544, no. 7649 (2017): 223–26, https://doi.org/10.1038/nature21728.

153 **A lot of executives:** A McKinsey & Company study of more than a thousand board members and C-suite executives from corporations around the world found that 86 percent thought a longer time horizon for decision making would improve their company's performance, with stronger financial returns and increased innovation. The results of this survey conducted in 2013 by the Canadian Pension Plan Investment Board and McKinsey are described in Dominic Barton and Mark Wiseman, "Investing for the Long Term," McKinsey & Company, December 2014, https://www.mckinsey.com/industries/private-equity-and-principal-investors/our-insights/investing-for-the-long-term.

153 **Ron Shaich, Panera Bread, and activist shareholders:** What I've written here is based on my interview with Shaich. This article expounds on that discussion: Craig Giammona, "Panera CEO Knocks Wall Street Culture That Celebrates Activists," Bloomberg.com, August 22, 2017, https://www.bloomberg.com/news/articles/2017-08-22/panera-ceo-knocks-wall-street-culture-that-celebrates-activists.

Another CEO of "fast casual" restaurant franchises I know, Greg Flynn of Flynn Holdings, has kept his company privately held over time instead of taking it public. His sense is that quarter-to-quarter earnings pressure would run against his long-range goals for the business.

Flynn's experience with private equity, however, was that it, too, had short time horizons—even if not tied to quarterly metrics like the stock market. In 2001, he brought in such investors to help grow the company and acquire more restaurants. Within a few years, the investors had an expectation of exiting the investment with high returns. Flynn could have sought yet a larger sum of private equity funding to buy out those investors, but instead he decided to go to a source of longer-term capital, the Ontario Teachers' Pension Plan. He had heard from a friend that Canadian institutional investors had been buying direct stakes in companies with prospects for long-term growth, instead of going through money managers, and it turned out they were willing to invest $300 million in Flynn Holdings. They now hold the majority stake in the company. For more on Flynn, see Amy Feldman, "The Super Sizer: How Greg Flynn Became America's Largest Restaurant Franchisee with $1.9B Revenues," *Forbes*, August 24, 2016, https://www.forbes.com/sites/amyfeldman/2016/08/24/the-super-sizer-how-greg-flynn-became-americas-largest-restaurant-franchisee-with-1-9b-revenues/#1639d56624b9. For more on the strategy that both Flynn and Shaich pursued, see Joseph Cotterill, "'Direct Investors' a Growing Force in Private Markets," *Financial Times*, June 16, 2015, https://www.ft.com/content/5b002968-1404-11e5-9bc5-00144feabdc0.

154 **JAB:** JAB Holding Company's current leaders are now trying to atone for its racist past and Nazi origins. See Katrin Bennhold, "Nazis Killed Her Father. Then She Fell in Love with One," *The New York Times*, June 14, 2019, https://www.nytimes.com/2019/06/14/business/reimann-jab-nazi-keurig-krispy-kreme.html.

154 **Jamie Dimon:** See Tara Lachapelle, "Jamie Dimon Forgot to Mention Mergers Are Part of the Problem," Bloomberg.com, April 5, 2017, https://www.bloomberg.com/gadfly/articles/2017-04-05 /jamie-dimon-forgot-to-mention-mergers-are-part-of-the-problem. See also Dimon's 2016 annual shareholder letter to JPMorgan Chase: https://www.jpmorganchase.com/corporate/investor-relations /document/ar2016-ceolettershareholders.pdf.

154 **what are known as benefit corporations:** More information about these corporations can be found at http://benefitcorp.net/.

155 **Today, companies' R&D investment has shifted:** Robert D. Atkinson and Luke A. Stewart, "University Research Funding: The U.S. Is Behind and Falling," Information Technology & Innovation Foundation, 2011, https://itif.org/publications/2011/05/19/university-research-funding-united-states -behind-and-falling; OECD, *OECD Science, Technology and Industry Outlook 2014* (Paris: OECD Publishing, 2014), https://doi.org/10.1787/sti_outlook-2014-en; President's Council of Advisors on Science and Technology (PCAST), "Report to the President. Transformation and Opportunity: The Future of the U.S. Research Enterprise," Executive Office of the President, 2012, https://obamawhitehouse .archives.gov/sites/default/files/microsites/ostp/pcast_future_research_enterprise_20121130.pdf.

156 **Lynn Stout . . . "Consider the dilemma a shareholder faces":** Lynn A. Stout, "The Corporation as Time Machine: Intergenerational Equity, Intergenerational Efficiency, and the Corporate Form," *Seattle University Law Review* 38, no. 2 (2015), https://digitalcommons.law.seattleu.edu/sulr/vol38/iss2/18/.

156 **the average holding time for a stock is just a few months:** This high turnover has been documented by Stout, "The Corporation as Time Machine," and is supported by New York Stock Exchange data.

156 **Andrew Haldane:** Haldane was executive director of financial stability at the Bank of England as of this writing. See Andrew Haldane, "Patience and Finance," Oxford China Business Forum speech, September 9, 2010, https://www.bis.org/review/r100909e.pdf; and Andrew Haldane and Richard Davies, "The Short Long," in *New Paradigms in Money and Finance?* 29th Société Universitaire Européene de Recherches Financières Colloquium, May 2011, https://www.bankofengland.co.uk /-/media/boe/files/speech/2011/the-short-long-speech-by-andrew-haldane.

157 **financial transaction tax:** American Prosperity Project, "A Nonpartisan Framework for Long-Term Investment," Aspen Institute report, 2016, https://www.aspeninstitute.org/programs/business -and-society-program/american-prosperity-project/. Another option is to make the capital gains tax more targeted at rewarding truly long-term investment. Currently, the capital gains tax is binary— higher taxes imposed on those who invest less than a year, and lower on those who invest a year or more. It does not offer greater reward for people who hold on to stocks for five years as opposed to one. The IRS rules could better reward people for investing over the course of years, particularly in small businesses, and penalize them for holding stock for less than a year.

158 **Dominic Barton of McKinsey has suggested a reform:** See Dominic Barton, "Capitalism for the Long Term," *Harvard Business Review*, March 2011, https://hbr.org/2011/03/capitalism-for-the-long-term.

158 **long-term stock exchange:** Alexander Osipovich and Dennis K. Berman, "Silicon Valley vs. Wall Street: Can the New Long-Term Stock Exchange Disrupt Capitalism?" *The Wall Street Journal*, October 16, 2017, https://www.wsj.com/articles/silicon-valley-vs-wall-street-can-the-new-long-term -stock-exchange-disrupt-capitalism-1508151600?mod=e2tw.

CHAPTER 6. THE GLITTER BOMB

159 **Wes Jackson and perennial grains:** My account was informed by interviews with multiple scientists and farmers, and site visits to the Land Institute in 2016. The following studies and articles give more context on perennial grains and informed my inquiry: J. D. Glover et al., "Increased Food and Ecosystem Security via Perennial Grains," *Science* 328, no. 5986 (2010): 1638–39, https://doi.org /10.1126/science.1188761; Stan Cox, "Ending 10000 Years of Conflict Between Agriculture and Nature," 2008, Science in Society Archive, http://www.i-sis.org.uk/Ending10000YearsOfConflict .php; Timothy E. Crews et al., "Going Where No Grains Have Gone Before: From Early to Mid-Succession," *Agriculture, Ecosystems & Environment* 223 (2016): 223–38, https://doi.org/10.1016 /j.agee.2016.03.012; Jerry D. Glover, Cindy M. Cox, and John P. Reganold, "Future Farming: A Return to Roots?" *Scientific American* 297, no. 2 (2007): 82–89, https://doi.org/10.1038/scientifi camerican0807-82; Daniel A. Kane, Paul Rogé, and Sieglinde S. Snapp, "A Systematic Review of Perennial Staple Crops Literature Using Topic Modeling and Bibliometric Analysis," *PLOS One* 11, no. 5 (2016), https://doi.org/10.1371/journal.pone.0155788; I. Lewandowski, "The Role of Perennial Biomass Crops in a Growing Bioeconomy," in *Perennial Biomass Crops for a Resource-Constrained*

World, ed. Susanne Barth et al. (Cham, Switzerland: Springer Nature, 2016), pp. 3–13, https://doi
.org/10.1007/978-3-319-44530-4_1.

160 **the Dust Bowl:** For more on the incredible history of the era, I recommend Timothy Egan, *The
Worst Hard Time: The Untold Story of Those Who Survived the Great American Dust Bowl* (Boston:
Houghton Mifflin, 2013).

163 **Danny Hillis:** My interviews with Hillis, previously described, and with Stewart Brand informed
this account. See also Stewart Brand, *The Clock of the Long Now: Time and Responsibility: The Ideas
Behind the World's Slowest Computer* (New York: Basic Books, 2008).

164 **Rachel Sussman:** Rachel Sussman, *The Oldest Living Things in the World* (Chicago: University of
Chicago Press, 2014).

165 **David Rosenthal and LOCKSS (Lots of Copies Keep Stuff Safe):** I interviewed Rosenthal on several occasions, including for my article "The Race to Preserve Disappearing Data," *The Boston Globe*,
May 17, 2015, https://www.bostonglobe.com/ideas/2015/05/16/the-race-preserve-disappearing
-data/0KPHAx5iK6jaLIvWQqIl4O/story.html.

167 **Gregg Popovich's practices and the NBA's reactions:** My description was informed by an interview
with one of Coach Pop's former colleagues and by the following articles. For more on the culture of
the NBA, see Sam Hinkie's resignation letter at https://www.espn.com/pdf/2016/0406/nba_hinkie
_redact.pdf. See also Todd Whitehead, "NBA Teams Are Resting Players Earlier and Earlier,"
FiveThirtyEight, March 31, 2017, https://fivethirtyeight.com/features/nba-teams-are-resting-players
-earlier-and-earlier/; Kevin Arnovitz, "The NBA's Culture Warriors," ESPN Internet Ventures,
May 23, 2017, http://www.espn.com/nba/story/_/page/presents19431278/the-warriors-spurs-fighting
-soul-game; Chris Herring, "The Spurs Have Evolved Their Way to Greatness," FiveThirtyEight,
May 12, 2017, https://fivethirtyeight.com/features/the-spurs-have-evolved-their-way-to-greatness/;
Ramona Shelburne, "Adam Silver: Resting Star Players 'a Significant Issue for the League,'" ESPN
Internet Ventures, March 20, 2017, http://www.espn.com/nba/story/_/id/18962901/resting-star
-players-significant-issue-league; and Mark C. Drakos et al., "Injury in the National Basketball
Association: A 17-Year Overview," *Sports Health: A Multidisciplinary Approach* 2, no. 4 (2010): 284–90,
https://doi.org/10.1177/1941738109357303.

169 **a perennial grain that resembles wheat:** The grain has been trademarked by the Land Institute and
is called Kernza.

170 **New Guinea and casuarina trees:** Jared Diamond, *Collapse: How Societies Choose to Fail or Succeed*
(New York: Penguin, 2011).

170 **Toyota Motor Corporation:** The described meetings at Toyota took place in Japan in March 2017. I
was also shown a confidential slide deck demonstrating the phenomenon I describe.

171 **lentils and vaccination in Rajasthan:** A. V. Banerjee et al., "Improving Immunisation Coverage in
Rural India: Clustered Randomised Controlled Evaluation of Immunisation Campaigns with and
Without Incentives," *The BMJ* 340 (2010): c2220, https://doi.org/10.1136/bmj.c2220.

172 **Andrew Lo and megafunds:** What I describe was informed by an interview with Lo in 2017 and by
Jose-Maria Fernandez, Roger M. Stein, and Andrew W. Lo, "Commercializing Biomedical Research Through Securitization Techniques," *Nature Biotechnology* 30, no. 10 (2012): 964–75, https://
doi.org/10.1038/nbt.2374.

175 **Raymond Orteig put up a prize:** Tim Brady, "The Orteig Prize," *Journal of Aviation/Aerospace Education & Research* 12, no. 1 (2002), https://doi.org/10.15394/jaaer.2002.1595.

176 **the Longitude Prize:** Dava Sobel, *Longitude: The True Story of a Lone Genius Who Solved the Greatest
Scientific Problem of His Time* (New York: Bloomsbury, 2007).

177 **DARPA, held another kind of competition:** Alex Davies, "An Oral History of the Darpa Grand
Challenge, the Grueling Robot Race That Launched the Self-Driving Car," *Wired*, August 3, 2017,
https://www.wired.com/story/darpa-grand-challenge-2004-oral-history/.

177 **more thoughts on innovation prizes:** Prize competitions ideally imagine the feat to be accomplished, but not the way of doing it—crossing a desert quickly and safely with an autonomous vehi-
cle or coming up with a treatment for Lou Gehrig's disease, but not the specific technology or type
of cure. That way, the competition can leave open the possible ways to solve the problem, says
Karim Lakhani, a Harvard Business School professor who has studied the contemporary innovation
prize. Designing a prize this way elicits ideas that come from far afield—for example, from a clock-
maker instead of an astronomer. There is an emerging literature on the resurgence of innovation
prizes, notably: "And the Winner Is . . . : Capturing the Promise of Philanthropic Prizes," McKinsey
& Company, July 2009, accessed at https://www.mckinsey.com/~/media/mckinsey/industries/social
%20sector/our%20insights/and%20the%20winner%20is%20philanthropists%20and%20govern

ments%20make%20prizes%20count/and-the-winner-is-philanthropists-and-governments-make -prizes-count.ashx; and Fiona Murray et al., "Grand Innovation Prizes: A Theoretical, Normative, and Empirical Evaluation," *Research Policy* 41, no. 10 (2012): 1779–92, https://doi.org/10.1016/j.respol .2012.06.013.

One reason prizes can work to overcome bias for the present is that competition with others, in some instances, motivates people to value deferred rewards more highly. Nicholas Christenfeld, a psychology professor at the University of California at San Diego, has demonstrated this phenomenon in studies of dating competitions and races. Social competition, in his estimation, motivates people to value something they might not otherwise value and to persevere to get it over time. This kind of competition can be detrimental if the object of desire is a date with a human being you don't actually like that much, but potentially positive if it motivates you to solve societal or scientific problems. (The dollar amount needs to be proportionate to the kind of effort required—a $2,000 prize would likely not have inspired private citizens to take a spaceship into orbit.)

In some cases, commercial opportunities that arise from an innovation prize competition far surpass the dollar value of the prize purse, because a market emerges along the way for the technologies. Blair Effron, the banker behind Centerview Partners, has argued that such prizes are preferable to patents in many cases because they not only immediately reward invention and research discoveries, but encourage the widespread sharing and use of technologies by more people—effectively a down payment on future innovations.

When prize competitions are at their best, they capture people's imaginations to make the daunting seem possible. They take a problem of concern and recast it—imagining it as being solved in some future rather than intractable, and motivating people to project themselves into the winning scenario by luring them with a prize. John F. Kennedy's 1961 pledge to put a man on the moon within the decade was not a prize competition, but similarly envisioned an audacious outcome, one that enraptured its own generation and also inspired future scientists and engineers to pursue their careers. It was also the race against the Soviet space program, which had launched Sputnik a few years earlier, that inspired the feat. Fierce competition frequently motivates people to overcome current obstacles—as it did in the 1990s in scientists' race to sequence the human genome, when one group wanted to put the information into the public domain for wide use and the other to make it the province of private companies.

It's worth noting that prizes are also not the answer to all of society's needs for exploration and research. Not every discovery is something we can imagine. In fact, many life-changing technologies have arisen across history simply because people entertained their curiosity. The glittering skylines of Shanghai and New York, night trains through the desert, power drills, computers, and countless conveniences and technologies based on electricity rely on Michael Faraday's experiments with electromagnetic fields in the 1830s, once deemed frivolous. GPS would take us to the wrong place if not for physicists who drew on Einstein's theory of relativity, which he developed not with any future navigation system in mind but led by his curiosity about the nature of the universe and time. We can't just imagine innovation feats; we also have to indulge curiosity about the unknown. So much of what lies in the future suffers from being hard or even impossible to imagine, and yet it can change the world—by ushering in discovery or disaster.

CHAPTER 7. HELL OR HIGH WATER

181 **civilizations across history have fallen from lofty heights:** See Jared Diamond, *Collapse: How Societies Choose to Fail or Succeed* (New York: Penguin, 2011).

183 **the Green Diamond proposal and Kit Smith in Richland County, South Carolina:** I relied on a range of sources to reconstruct the events leading up to and following the battle over Green Diamond: transcripts, trial briefs and exhibits, and judicial opinions involved in the lawsuit and court proceedings; extensive interviews with relevant parties and external experts; local and national media coverage from the time period in question; government maps; and my own reporting and site visits. I am grateful for the many experts and sources who spent time educating me about the case, and in particular to Mullen Taylor, who spent an extraordinary amount of time and effort helping me excavate court documents and clarifying key issues. I am also grateful to Samantha Medlock, who first called my attention to what had happened in Richland County.

184 **Columbia Venture enlisted big political guns:** Michael Grunwald, "For S.C. Project, a Torrent of Pressure; Developer Wins Reprieve from FEMA on $4 Billion Project in Columbia Flood Plain," *The Washington Post*, July 13, 2001, http://www.highbeam.com/doc/1P2-469170.html?refid=easy_hf.

185 **As of 2016, the National Flood Insurance Program:** *GAO's 2017 High Risk Report: 34 Programs in Peril: Hearing Before the Committee on Oversight and Government Reform, House of Representatives, One Hundred Fifteenth Congress, First Session, February 15, 2017,* https://oversight.house.gov/hearing /gaos-2017-high-risk-report-34-programs-peril/.

185 **One house in Spring, Texas, documented by journalist Mary Williams Walsh:** Mary Williams Walsh, "A Broke, and Broken, Flood Insurance Program," *The New York Times,* November 4, 2017, https:// www.nytimes.com/2017/11/04/business/a-broke-and-broken-flood-insurance-program.html?smid =tw-share.

185 **More than thirteen hundred communities around the country:** The sources for this discussion are government data and fact sheets associated with the U.S. Federal Emergency Management Agency's Community Rating System, https://www.fema.gov/community-rating-system.

187 **Howard Kunreuther calculated in 2009:** Howard Kunreuther, Robert Meyer, and Erwann Michel-Kerjan, "Overcoming Decision Biases to Reduce Losses from Natural Catastrophes," Risk Management and Decision Processes Center, The Wharton School, University of Pennsylvania, 2009, http://opim.wharton.upenn.edu/risk/library/C2009_HK,RJM,EMK.pdf.

187 **A report commissioned in 2017 by the U.S. government:** Multihazard Mitigation Council (K. Porter, principal investigator), *Natural Hazard Mitigation Saves 2017 Interim Report: An Independent Study* (Washington, DC: National Institute of Building Sciences, 2017).

188 **Yet numerous experts today point out that it is a fallacy:** Joseph E. Stiglitz, Amartya Sen, and Jean-Paul Fitoussi, *Mismeasuring Our Lives: Why GDP Doesn't Add Up* (New York: New Press, 2010).

189 **former inhabitants of Easter Island:** See Diamond, *Collapse.*

189 **Tyler Cowen:** Tyler Cowen, *Stubborn Attachments: A Vision for a Society of Free, Prosperous, and Responsible Individuals* (San Francisco: Stripe Press, 2018).

191 **officials knew of the potential danger and scope of the epidemic:** Maria Cheng and Raphael Satter (Associated Press), "Emails Show the World Health Organization Intentionally Delayed Calling Ebola a Public Health Emergency," *Business Insider,* March 20, 2015, https://www.businessinsider.com /report-the-world-health-organization-resisted-declaring-ebola-an-international-emergency-for -economic-reasons-2015-3.

191 **compared this excuse to not calling the fire department when several houses are on fire:** Ibid.

191 **Barbara Tuchman:** Barbara W. Tuchman, *The March of Folly: From Troy to Vietnam* (New York: Random House Trade Paperbacks, 2014).

192 **history of Burroughs family and Burroughs & Chapin company:** My account was informed by interviews, court documents, and Susan Hoffer McMillan's history of Myrtle Beach, *Myrtle Beach and the Grand Strand* (Charleston, SC: Arcadia, 2004).

195 **the Takings Clause and Pennsylvania Coal:** Robert Brauneis, "'The Foundation of Our "Regulatory Takings" Jurisprudence': The Myth and Meaning of Justice Holmes's Opinion in Pennsylvania Coal Co. v. Mahon," *The Yale Law Journal* 106, no. 3 (1996): 613–702, https://doi.org/10.2307/ 797307.

196 **raising the costs of extreme weather events:** U.S. Department of Commerce and NOAA, "Billion-Dollar Weather and Climate Disasters: Overview," National Centers for Environmental Information, National Climatic Data Center, https://www.ncdc.noaa.gov/billions, accessed August 20, 2018. Global data from the World Meteorological Organization.

199 **"lust for power":** See Tuchman, *The March of Folly.*

199 **Mo Ibrahim Prize:** See http://mo.ibrahim.foundation/prize/.

200 **Citizens United:** *Citizens United v. Federal Election Commission,* U.S. Supreme Court, 2010, https:// www.supremecourt.gov/opinions/09pdf/08-205.pdf.

201 **At the city and state levels, however, movements are afoot:** Ashley Balcerzak, "Statehouses, Not Congress, Hosting Biggest Political Money Fights," *Time,* August 31, 2017, http://time.com /4922560/statehouses-congress-money-fights-campaign-finance/.

202 **Solon:** Debra Hamel, *Reading Herodotus: A Guided Tour through the Wild Boars, Dancing Suitors, and Crazy Tyrants of* The History (Baltimore: Johns Hopkins University Press, 2012).

202 **Graham Allison:** Robert O'Neill, "Teaching in Time: Using History to Teach Future Public Policy Practitioners," *Harvard Kennedy School Magazine,* 2017, https://www.hks.harvard.edu/more/alumni /hks-magazine/teaching-time-using-history-teach-future-public-policy-practitioners.

202 **Amanda Lynch and Siri Veland criticize the designation of our epoch as the Anthropocene:** Amanda H. Lynch and Siri Veland, *Urgency in the Anthropocene* (Cambridge, MA: MIT Press, 2018).

206 **"One entered the city like a god":** Vincent Scully, *American Architecture and Urbanism* (New York: Frederick A. Praeger, 1969), p. 143.

207 **The U.S. Supreme Court ultimately heard the case:** J. D. Echeverria, "Making Sense of *Penn Central*," *UCLA Journal of Environmental Law and Policy* 23, no. 2 (2005), https://escholarship.org/uc/item/0vz8057f.

207 **"Not only do these buildings and their workmanship":** Penn Central Transportation Co. et al. v. New York City et al., Supreme Court of the United States, 438 U.S. 104 (decided June 26, 1978), accessed at http://law2.umkc.edu/faculty/projects/ftrials/conlaw/penncentral.html.

209 **In October 2015, the South Carolina midlands:** U.S. Department of Commerce and NOAA, "Historic Flooding—October 1–5, 2015," National Weather Service, October 29, 2015, https://www.weather.gov/chs/HistoricFlooding-Oct2015.

CHAPTER 8. THE GAMES WE PLAY

212 **Georg Sieber and Situation 21 at the Munich Olympics:** My account of these events was informed by news reports, books, and notably Sarah Morris's brilliant documentary short *1972*. Unfortunately, I did not interview Sieber as I did most of the people described in this book, but I am grateful to Morris for her accomplished interview, which allowed me to observe him indirectly, and to Alexander Wolff for his reflective reporting on the Munich Olympics and Sieber's role. Key sources include: Aaron J. Klein and Mitch Ginsburg, *Striking Back: The 1972 Munich Olympics Massacre and Israel's Deadly Response* (New York: Random House Trade Paperbacks, 2007); Sarah Morris, dir., *1972* (Germany, 2008); "Horror and Death at the Olympics," *Time*, September 18, 1972; Alexander Wolff, "When the Terror Began," *Time*, August 25, 2002; Alexander Wolff, "In Reconstructing Munich Massacre, I Learned History Is Ever-Present," SI.com, July 26, 2012, https://www.si.com/more-sports/2012/07/26/munich-massacre-40-years-later.

213 **after 9/11 many companies began paying exorbitant premiums:** Robert J. Meyer and Howard Kunreuther, *The Ostrich Paradox: Why We Underprepare for Disasters* (Philadelphia: Wharton Digital Press, 2017).

214 **TEPCO and the risk of tsunami:** James M. Acton and Mark Hibbs, "Why Fukushima Was Preventable," Carnegie Endowment for International Peace, 2012, https://carnegieendowment.org/files/fukushima.pdf.

215 **In the 1960s, the engineer, Yanosuke Hirai, had insisted:** There were also cultural norms set at Onagawa, in part by Hirai, that helped make it safer during the 2011 event. See Airi Ryu and Najmedin Meshkati, "Culture of Safety Can Make or Break Nuclear Power Plants," *The Japan Times*, March 11, 2014, https://www.japantimes.co.jp/opinion/2014/03/14/commentary/japan-commentary/culture-of-safety-can-make-or-break-nuclear-power-plants/#.W3wv8dhKiso.

216 **When Seneca wrote *Natural Questions*:** Lucius Annaeus Seneca, *Naturales Quaestiones*, trans. Thomas H. Corcoran (Cambridge, MA: Harvard University Press, 1971).

217 **the "last mile" of vaccination:** D. S. Saint-Victor and S. B. Omer, "Vaccine Refusal and the Endgame: Walking the Last Mile First," *Philosophical Transactions of the Royal Society B: Biological Sciences* 368, no. 1623 (2013), https://doi.org/10.1098/rstb.2012.0148.

217 **California earthquake insurance trends:** See Meyer and Kunreuther, *The Ostrich Paradox*.

218 **Two ratings agencies, Moody's and S&P:** See Nate Silver, *The Signal and the Noise: The Art and Science of Prediction* (New York: Penguin, 2012); and *The Financial Crisis Inquiry Report: Final Report of the National Commission on the Causes of the Financial and Economic Crisis in the United States* (Washington, DC: Government Printing Office, 2011), pp. 147–48, https://www.gpo.gov/fdsys/pkg/GPO-FCIC/pdf/GPO-FCIC.pdf.

219 **Jim Mattis:** I interviewed Mattis and consulted media coverage of his philosophies for this account and quotations. Mattis's studies of the enemy have also been guided by history. Osama bin Laden and, more recently, ISIS, he told me, have used fear as Hitler did, even if it manifested differently. "They recognize the value of fear in getting what they want and achieving what they want," he said.

220 **Ernest May and Richard Neustadt:** Richard E. Neustadt and Ernest R. May, *Thinking in Time: The Uses of History for Decision Makers* (New York: Free Press, 1988).

222 **stone markers in Murohama and Aneyoshi:** OECD, "Preservation of Records, Knowledge and Memory Across Generations (RK and M): Markers—Reflections on Intergenerational Warnings in the Form of Japanese Tsunami Stones," Nuclear Energy Agency of the OECD (NEA), 2014, https://www.oecd-nea.org/rwm/docs/2014/rwm-r2014-4.pdf.

223 **Mattapoisett and its flood marker project:** This account was informed by firsthand interviews, site visits, Environmental Protection Agency documents, and the archives of the Mattapoisett Historical Society. I am grateful to Jeri Weiss for pointing me to this community effort and for her expertise.

226 **Nassim Taleb:** Nassim Nicholas Taleb, *The Black Swan: The Impact of the Highly Improbable* (New York: Penguin, 2008).

226 **"Unprecedented is increasingly the norm":** Katharine Mach and Miyuki Hino, "What Climate Scientists Want You to See in the Floodwaters," *The New York Times*, September 2, 2017, https://www.nytimes.com/2017/09/02/opinion/climate-hurricanes-flooding-harvey.html.

228 **Peter Schwartz:** Peter Schwartz, *The Art of the Long View: Planning for the Future in an Uncertain World*, 2nd ed. (New York: Doubleday Currency, 1996).

228 **Peter Railton:** I interviewed Railton in September 2017. His perspective is corroborated by studies of response to the cone of uncertainty in hurricane forecasts, elaborated upon in the notes for chapter 1. See also Martin E. P. Seligman et al., *Homo Prospectus* (Oxford: Oxford University Press, 2017).

229 **Thomas Suddendorf:** From my interview of Suddendorf, noted in chapter 1. Suddendorf believes it pays to be cautious when presenting people with vivid and negative future scenarios. If the future looks too dire, it can make current choices seem meaningless. (For example, research shows that smokers that are pointed toward the risks of early death tend to smoke more immediately afterward.) On the other hand, if it seems like a person can easily shape the future—and they see that it's possible to avoid the worse consequences—it can embolden her to value the future. "Optimism is a strategy that can make the future better because it makes you want to work to make it so," he told me.

229 **Dr. Smith Dharmasaroja and tsunami warning:** Robert J. Meyer and Howard Kunreuther, *The Ostrich Paradox: Why We Underprepare for Disasters* (Philadelphia: Wharton Digital Press, 2017).

230 **Pablo Suarez and disaster games:** My interviews with Suarez from 2014 to 2017 inform this account, which included playing his games as well as observing a range of people playing them. I also interviewed aid experts who have witnessed his games being played globally. See also Pablo Suarez, "Rethinking Engagement: Innovations in How Humanitarians Explore Geoinformation," *ISPRS International Journal of Geo-Information* 4, no. 3 (2015): 1729–49, https://doi.org/10.3390/ijgi4031729.

232 **Peter Perla and war games:** I interviewed Perla and consulted his writings on this topic, particularly the following source, which includes the history of Naval College war games: Peter P. Perla, *The Art of Wargaming: A Guide for Professionals and Hobbyists* (Annapolis, MD: Naval Institute Press, 1990).

234 **the attack on Pearl Harbor, while its timing was a surprise, was actually expected:** Steven A. Sloman and Philip Fernbach, *The Knowledge Illusion: Why We Never Think Alone* (New York: Riverhead Books, 2017).

234 **Pentagon war games:** My interviews with Frost and McCown, as well as corroborating interviews with a confidential source, informed this account.

236 **"One thing a person cannot do":** Thomas Schelling, "The Role of Wargames and Exercises," in *Managing Nuclear Operations*, ed. Ashton B. Carter, John D. Steinbruner, and Charles A. Zraket (Washington, DC: Brookings Institution Press, 1987), pp. 426–44.

236 **Dick Cheney and Dark Winter game:** Jane Mayer, *The Dark Side: The Inside Story of How the War on Terror Turned into a War on American Ideals* (New York: Doubleday, 2008).

237 **Larry Susskind and community climate games:** I am grateful to Danya Rumore who spent time informing me about the details of the climate games and enriched my view of what games can accomplish. For the research findings, see Danya Rumore, Todd Schenk, and Lawrence Susskind, "Role-Play Simulations for Climate Change Adaptation Education and Engagement," *Nature Climate Change* 6, no. 8 (2016): 745–50, https://doi.org/10.1038/nclimate3084.

238 **"When we play, we also have a sense of urgent optimism":** See Jane McGonigal's 2010 TED talk "Gaming Can Make a Better World," for this quotation, and her book: *Reality Is Broken: Why Games Make Us Better and How They Can Change the World* (New York: Penguin Press, 2011).

CHAPTER 9. THE LIVING CROWD

240 **Maureen Kaplan and the nuclear waste disposal question:** My account of the Buffalo meeting and Maureen Kaplan's investigation was based on interviews with Kaplan; the extensive report that emerged from the meetings in the early '90s, which described both the process and the findings; as well as several news articles from the time, notably Kathleen M. Trauth, Stephen C. Hora, and Robert V. Guzowski, *Expert Judgment on Markers to Deter Inadvertent Human Intrusion into the Waste Isolation Pilot Plant*, Sandia Report (Washington, DC: United States Department of Energy, 1993), https://prod.sandia.gov/techlib-noauth/access-control.cgi/1992/921382.pdf; and Maureen F. Kaplan

and Mel Adams, "Using the Past to Protect the Future: Marking Nuclear Waste Disposal Sites," *Archaeology* 39, no. 5 (1986): 51–54, http://www.jstor.org/stable/41731805.

242 **technology called CRISPR:** I have written about this topic in "Should We Engineer Future Humans?" *New America Weekly*, December 17, 2015, https://www.newamerica.org/weekly/104/should-we-engineer-future-humans/.

244 **In 1980, the U.S. government had asked several technical experts:** These meetings and the report that ensued are documented here: Battelle Memorial Institute and U.S. Department of Energy, *Reducing the Likelihood of Future Human Activities That Could Affect Geologic High-Level Waste Repositories: Technical Report* (Columbus, OH: Office of Nuclear Waste Isolation, Battelle Memorial Institute, 1984). See also Scott Beauchamp, "How to Send a Message 1,000 Years to the Future," *The Atlantic*, February 24, 2015, https://www.theatlantic.com/technology/archive/2015/02/how-to-send-a-message-1000-years-to-the-future/385720/; and Juliet Lapidos, "Atomic Priesthoods, Thorn Landscapes, and Munchian Pictograms: How to Communicate the Dangers of Nuclear Waste to Future Civilizations," *Slate*, November 16, 2009, http://www.slate.com/articles/health_and_science/green_room/2009/11/atomic_priesthoods_thorn_landscapes_and_munchian_pictograms.html.

244 **Thomas Sebeok:** See the Battelle report and Lapidos, "Atomic Priesthoods, Thorn Landscapes, and Munchian Pictograms," both cited in the preceding note.

244 **Decades later, in 2014, another group:** Organisation for Economic Co-operation and Development, *Radioactive Waste Management and Constructing Memory for Future Generations: Proceedings of the International Conference and Debate: 15–17 September 2014*, 2015, https://www.oecd-nea.org/rwm/pubs/2015/7259-constructing-memory-2015.pdf.

244 **engineers in Finland have been constructing:** Andrew Curry, "What Lies Beneath," *The Atlantic*, October 2017, https://www.theatlantic.com/magazine/archive/2017/10/what-lies-beneath/537894/.

245 **Yucca Mountain:** For more on this history, see William M. Alley and Rosemarie Alley, *Too Hot to Touch: The Problem of High-Level Nuclear Waste* (Cambridge, England: Cambridge University Press, 2013).

246 **it took less than seven decades from the first airplane flight:** Shermer, a science columnist and self-described skeptic, made the average life span of civilization calculations himself. Estimates of the average life span can vary among scholars according to the point at which they deem a civilization as beginning or over. Michael Shermer, "Why ET Hasn't Called," *Scientific American*, August 2002, available at https://michaelshermer.com/2002/08/why-et-hasnt-called/.

246 **The amount of progress and disruption:** Ray Kurzweil, "The Law of Accelerating Returns," in *Alan Turing: Life and Legacy of a Great Thinker*, ed. Christof Teuscher (Berlin, Heidelberg, and New York: Springer, 2004).

246 **"future shock"—a kind of cultural disorientation:** Alvin Toffler, *Future Shock* (New York: Random House, 1970).

247 **John Maynard Keynes:** John Maynard Keynes, *A Tract on Monetary Reform* (London: Macmillan, 1923).

247 **Looking back, someone might have shot Hitler's grandparents:** For more on this line of reasoning, see Tyler Cowen, *Stubborn Attachments: A Vision for a Society of Free, Prosperous, and Responsible Individuals* (San Francisco: Stripe Press, 2018).

248 **"social discounting":** For further discussion, see the following sources: Will Oremus, "How Much Is the Future Worth? The Arcane, Fascinating Academic Debate That Helps Explain Why We Didn't Prepare for Hurricane Harvey," *Slate*, September 1, 2017, http://www.slate.com/articles/technology/future_tense/2017/09/how_social_discounting_helps_explain_why_we_don_t_prepare_for_disasters.html; Talbot Page, "Discounting and Intergenerational Equity," *Futures* 9, no. 5 (1977): 377–82, https://doi.org/10.1016/0016-3287(77)90019-2; Hal R. Varian, "Recalculating the Costs of Global Climate Change," *The New York Times*, December 14, 2006, https://www.nytimes.com/2006/12/14/business/14scene.html; Jonathan Orlando Zaddach, "Consumer Sovereignty vs. Intergenerational Equity: An Overview of the Stern-Nordhaus Debate," in *Climate Policy Under Intergenerational Discounting* (Wiesbaden, Germany: Springer Gabler, 2015), pp. 11–16, https://doi.org/10.1007/978-3-658-12134-1_3.

249 **Consider a hypothetical question:** See Cowen, *Stubborn Attachments*.

251 **concern for future generations is a near-universal human value:** A thorough discussion of this point can be found in Edith Brown Weiss, *In Fairness to Future Generations: International Law, Common Patrimony, and Intergenerational Equity* (Tokyo: United Nations University, 1989); and Mary Christina Wood, *Nature's Trust: Environmental Law for a New Ecological Age* (New York: Cambridge University Press, 2014).

251 "not only between those who are living": Edmund Burke, *Reflections on the Revolution in France* (1790), in *Select Works of Edmund Burke: A New Imprint of the Payne Edition*, foreword by Francis Canavan (Indianapolis: Liberty Fund, 1999).

251 Edith Brown Weiss: I interviewed Weiss in 2017. See also Weiss, *In Fairness to Future Generations*.

251 Thomas Jefferson and Theodore Roosevelt: See Wood, *Nature's Trust*, p. 264.

252 Michael Blumm has documented how this doctrine: Michael C. Blumm and Rachel D. Guthrie, "Internationalizing the Public Trust Doctrine: Natural Law and Constitutional and Statutory Approaches to Fulfilling the Saxion Vision" (April 20, 2011), *University of California Davis Law Review* 44 (2012); Lewis & Clark Law School Legal Studies Research Paper No. 2011-12, available at https://ssrn.com/abstract=1816628.

252 Customary land law in Ghana: See Weiss, *In Fairness to Future Generations*, p. 20, for a discussion.

252 idea that we should act in deference to future generations also permeates: See Wood, *Nature's Trust*, p. 265, and Weiss, *In Fairness to Future Generations*, pp. 18–19.

253 In letters he exchanged with a UCLA composer and professor: The letters exchanged between my great-grandfather K. V. Ramachandran and the composer and ethnomusicologist Colin McPhee can be found in the Colin McPhee Collection in the Ethnomusicology Archive at the University of California, Los Angeles: https://oac.cdlib.org/findaid/ark:/13030/kt9c6029pt/entire_text/.

255 Mary Wood: I interviewed Wood in 2017. See also her book *Nature's Trust*.

256 Richard Zeckhauser and Larry Summers: I interviewed Zeckhauser and had a brief conversation with Summers about this matter. See also Lawrence Summers and Richard Zeckhauser, "Policymaking for Posterity," *Journal of Risk and Uncertainty* 37, no. 2/3 (2008): 115–40, http://www.jstor.org/stable/41761455.

258 Baja California Sur lobster fishing cooperative: My descriptions are based on extensive interviews and research I conducted in 2007 and 2008 in Baja California Sur and with Mexican fisheries officials and scientists, while a graduate student at Harvard's Kennedy School of Government. Some of the content was previously compiled in a research paper I coauthored with Andres Schabelman, "The Political Implications of Fisheries Certification: Lessons from Baja California Sur" (April 1, 2008). My account here was also informed by Bonnie J. McCay et al., "Cooperatives, Concessions, and Co-Management on the Pacific Coast of Mexico," *Marine Policy* 44 (2014): 49–59, https://doi.org/10.1016/j.marpol.2013.08.001.

    My reflection on the fishery as a shared heirloom came more recently. My study of other societies and the scholarship of others has bolstered my view that it is possible to organize collectively to steward resources as heirlooms as the Baja California Sur lobster fishing communities have done. In *Collapse*, Jared Diamond analyzes the Tokugawa society that thrived in Japan for 250 years despite severe resource challenges. He examines the success of that civilization that endured from the early seventeenth century until 1867, as compared with other societies that collapsed when faced with similar existential threats such as deforestation. Among the choices made by Tokugawa society that Diamond describes are what I would call ways of making formal the idea of collective heirlooms. The government gave out long-term contracts to villages and merchants to manage forest land, giving them a stake in the future of the forests—similar to what the Gulf of Mexico red snapper fishery did with its catch shares. Peasants and villagers of Tokugawa Japan hoped to pass their land on to their heirs, expecting their children to inherit the rights to use the land like the Baja California Sur lobster fishermen. (The elites also had a personal interest in exercising foresight, because they expected their families to continue to control Japan into the foreseeable future.)

259 Stewart Brand: I interviewed Brand in spring 2016. See also his book *The Clock of the Long Now: Time and Responsibility: The Ideas Behind the World's Slowest Computer* (New York: Basic Books, 2008).

259 Bosco Che Suona: Christopher Livesay, "In the Italian Alps, Stradivari's Trees Live On," NPR, December 6, 2014, https://www.npr.org/sections/deceptivecadence/2014/12/05/368718313/in-the-italian-alps-stradivaris-trees-live-on.

261 intergenerational equity should be enforced more aggressively: Edith Brown Weiss, "Ensuring Fairness to Future Generations," in *Intergenerational Equity in Sustainable Development Treaty Implementation*, ed. Marcel Szabó and Marie-Claire Cordonier Segger (New York: Cambridge University Press, 2019).

261 a groundbreaking lawsuit is making its way through the courts: This legal effort is being organized by the Oregon group Our Children's Trust: https://www.ourchildrenstrust.org. See also Wood, *Nature's Trust*, p. 228.

261 The Supreme Court of India has made two decisions: Weiss, "Ensuring Fairness to Future Generations."

263 **if the current medical progress trajectory continues:** See Lapidos, "How to Communicate the Dangers of Nuclear Waste to Future Civilizations."

266 **a national movement of teenagers asking parents to sign contracts pledging:** This effort is known as the Parents Promise to Kids. For more, see the website at https://www.parentspromisetokids.org.

CODA: HOPE FOR A RECKLESS AGE

270 **Paul Bain's research:** As noted in the text, I take such studies with a grain of salt. Nevertheless, the evidence on the power of agency in imagining the future and also on how dystopic scenarios demotivate us from using foresight, documented in chapters 1 and 8 and their endnotes, further support the conclusions of this research, as do my interviews with Peter Railton, Tracy Gleason, Thomas Suddendorf, and others.

Specific studies from Bain are Paul G. Bain et al., "Collective Futures: How Projections About the Future of Society Are Related to Actions and Attitudes Supporting Social Change," *Personality and Social Psychology Bulletin* 39, no. 4 (2013): 523–39, https://doi.org/10.1177/0146167213478200; and Paul G. Bain et al., "Promoting Pro-Environmental Action in Climate Change Deniers," *Nature Climate Change* 2, no. 8 (2012): 600–603, https://doi.org/10.1038/nclimate1532.

271 **This research is hardly the final word:** On this topic, see my article "We Are the Future," *Slate*, September 14, 2017, http://www.slate.com/articles/technology/future_tense/2017/09/how_to_create _a_future_where_people_fight_climate_change.html.

Index